ELIN ANDREASSEN   HEIN B. BJERCK   BJØRNAR OLSEN

PERSISTENT MEMORIES
Pyramiden – a Soviet mining town in the High Arctic

tapir academic press

Contents

Encounter: the abandoned town.................11

Listening to things: fieldwork and representation 16

A Soviet mining town 28

Building a non-place..................................57

On entering our home in the abandoned Hotel Tulip on the third evening 69

Making a home... 111

A ruin in the true sense of the word 138

Persistent memories 152

Pyramiden in Svalbard 160

Pyramiden biography............................... 169

The last man 184

Endnotes.. 186

Bibliography 190

Photo index 195

PERSISTENT MEMORIES
Pyramiden – a Soviet mining town in the High Arctic

Encounter: the abandoned town

In 1998 the Russian mining company Trust Arcticugol (Arctic Coal Company) decided to end its more than 50 years of continuous activity in Пирамида (Pyramiden),[1] Svalbard. During that spring and summer all inhabitants deserted the mining town, leaving behind a site devoid of humans but still filled with all the paraphernalia constituting a modern city. The original plan for abandonment also included the dismantling and removal of equipment and infrastructure, stripping the town of all useful as well as environmentally threatening materials. However, apart from some superficial gestures to fulfil this goal, such as removing (but storing) fluorescent tubes, plugs and toilet seats, people and most things parted that summer, leaving the well-equipped Pyramiden more or less to its own destiny.[2]

When we arrived in 2006 little seemed to have changed. Buildings, walkways, monuments, piers, and various mining facilities gave face to a town stubbornly clinging to the barren ground. During the coming days we explored apartments, offices, workshops, mess halls, bars, the school, the hospital, and other buildings, all of which had been left there with most of their things in place. The abundant presence of the latter triggered a feeling of something postponed or put on hold rather than deserted: a Soviet town seemingly frozen in time. In the Cultural Palace most of the 60,000 books were still on shelves in the library, dusty musical instruments awaited players in the music studio, and the basketball court and movie theatre both lay ready-equipped under a blanket of seemingly temporary silence. Bottles and glasses may sit empty at the hotel bar, but otherwise time seemed to have stopped at a moment just after closing time. And at the museum next door, the Pyramiden exhibits were still on display.

At the same time, however, Pyramiden was showing signs of weathering and decay. Beyond the immediate impression of standstill, processes of decay were wrinkling the town's body: doors were broken; offices had been vandalized; walls and concrete were decomposing; and paint and wallpaper had surrendered to gravity, forming new archaeological layers. Nature was intruding and mingling. Nestling gulls competed over cramped spaces on window ledges, while rivers and streams, once held firmly in check by dikes and dams, had reclaimed their original delta space overtaken by the town. In addition, tourists and stray visitors visited Pyramiden in increasing numbers, taking keepsakes from the perceived ghost town.

Listening to things: fieldwork and representation

In July 2006 we arrived in Pyramiden in order to carry out fieldwork. For the next long week, two archaeologists and an art photographer constituted the town's entire human population, lodging at the deserted Hotel Tulip as the first visitors in eight years.[3] We came prepared with numerous questions and concerns about everyday life in the inhabited town, about its abandonment and its subsequent post-human fate. Fieldwork, however, always involves negotiation between preconceptions and encountered realities, and it was soon evident that Pyramiden also wanted to have a say.

Still, one central objective remained: to explore to what extent an 'archaeology of the present' could inform us about how people lived and coped in this Soviet town. Despite being a thoroughly modern community, there are few written records of everyday life in Pyramiden.[4] Thus, in this sense at least, we were encountering what may be described as a 'pre-historic'[5] society.

This anachronism notwithstanding, crucial to our approach is an appreciation of the social and historical importance of things, also as a source to understand modern literary societies. Written accounts are immensely important but are often selective, clustering around particular themes, events and persons considered significant. In contrast, things faithfully remember also the ineffable and those forgotten in 'talkative history',[6] giving face to the ordinary and the leftover. Being persistent and democratic, things allow for the marginal and dismissed to become part of our historical memory.

Partly preconceived, and partly invoked as we came to know Pyramiden, another issue of concern that became central was with Pyramiden's post-human biography and the way the site offered itself for more general reflection on things, heritage and history. To be confronted with a modern town filled with objects and yet devoid of what is normally conceived of as its primary constituents provokes reflections that hardly can be evoked voluntarily. Our everyday dealings with things mostly take place in a mode of inconspicuous familiarity; unless broken, interrupted or missing, things often pertain to a kind of shyness. Being at the same time 'the most obvious and the best hidden',[7] they largely escape our attention. Encountering a post-human Pyramiden, this mode of inconspicuous familiarity becomes lost or disturbed, and our relations with this 'new' material world are somehow uncomfortably disentangled. Not only do things become more present, more manifest, but in some ways they also become more pestering and disquieting. Things suddenly 'appear' to us in ways never noticed previously, exposing some of their own unruly 'thingness'.

Our encounter with the decaying Soviet town of Pyramiden on Norwegian Svalbard also triggered reflections on its value as heritage. Heritage is normally (and legally) conceived of as something 'old' and is also embedded in discourses of aesthetics and identity. How does a modern site such as Pyramiden, a decaying industrial ruin, fit into the common tropes of heritage and the current political economy of past? How does its conspicuous location in pristine Arctic nature, a setting infused with its own aesthetic tropes, impact on the way we conceive and value it? Further, if conceived as heritage and a site of remembrance, in what ways are memories preserved and disclosed, thus recalling the life of the Russians and Ukrainians who lived there?

Our fieldwork was preliminary and experimental, attempting to grasp and *sense* the place. The first requirement was simply to be present, to become embedded. After the ship which had transported us had left there was no escape, no temporarily post-work relief from the object of study, thus making experiencing of the place utterly imperative. The methodology applied was simple, low-tech and intuitive. The main modes of documentation comprised photography and taking notes and making sketches in diaries.

We first explored the town and the surrounding areas to familiarize ourselves with the site and then selected places and buildings we considered significant and could readily access for more detailed studies. Our permission to investigate included all buildings we could enter without having to break into them – in other words, where doors or other entrances were open.[8]

As this book make manifest, photography was crucial to our approach. Encountering things in their tacit material richness produces a 'topographic' experience that is both immediate and complex. At each moment, and at each place, one is exposed to a multitude of sensations that are hard to express or mediate through ordinary narrative accounts.[9] In our work photography became essential to grasp and mediate our Pyramiden experience.

Our commitment to photographic work is not primarily due to its superiority as a technical means of documentation, nor do we think of the role of images to be subservient to the text, in other words, functioning mainly as illustrations supporting written statements. When pictures outnumber text in this book it is not in order for it to be 'richly illustrated' but to allow things to speak through their own associative appearances. As noted by archaeologist Michael Shanks, 'photographs can introduce the heterogeneous and ineffable into discourse, that richness and detail in every photography which lies outside the categories and schemes of discourse'.[10] Thus, rather than disseminating our research in compliance with formats and genres of scientific prose where imagery only holds a secondary value, we have chosen this more experimental format. Challenging the traditional hierarchy of text over images, this book is also an attempt to negotiate some of the limitations set by traditional scholarship and create reflection by transcending boundaries between research and art.[11]

A Soviet mining town

Pyramiden is situated in Mimerbukta at the base of Billefjorden in the archipelago of Svalbard in the high north Atlantic Arctic (78°40´N 17°20´E). The closest settlements are the towns of Longyearbyen (Norwegian) and Barentsburg (Russian), located respectively 75 km and 100 km to the south-west. These are the two main remaining settlements on Svalbard, with a present population of *c.*2000 and *c.*500 inhabitants respectively.

The presence of both Norwegian and Russian settlements at Svalbard reflects the rather intricate political conditions that the islands are subject to. Although the adopted Svalbard Treaty (1920) assigns Norway 'full and absolute sovereignty' over the islands, it also gives the signature states equal rights to Svalbard's natural resources.[12] Since World War II only one of these states, the USSR, and now the Russian Federation, has made significant use of these rights by exploiting the huge coal reserves found on Svalbard. This means that throughout the entire Cold War Svalbard hosted both Norwegian and Soviet societies, thereby creating a rather unique geopolitical situation, not the least considering Norway's status as a faithful member of NATO since 1949.[13]

Pyramiden, as well as the still operating mining town of Barentsburg, is the property of the Russian state-owned coal company, Trust Arcticugol. While the company obtained its rights to the coalfields in Pyramiden prior to World War II, the official 'founding' of the town was associated with the start of the post-war rebuilding in 1946. Extensive operation of the mines also belonged to the post-war years and eventually led to regular shipments of coal. At its prime in the 1980s more than 1000 people resided in the town. These

inhabitants were recruited mainly from traditional mining communities in Ukraine and South Russia, primarily the cities of Donbass, Donetsk and Tula. Miners from these southern locations came to work in Pyramiden, which offers an average summer temperature of $c.5°C$, a winter average of $-12°C$ and permafrost reaching several hundred metres below the surface. The period of complete darkness during winter (approximately three months), however, is compensated for by four months of midnight sun[14]. With the exception of occasional helicopter visits, the town was isolated when the sea ice clogged the fjord from October to June.

However, as inhospitable as the climate and environmental conditions may seem, miners came to live in a modern town that provided them with most urban facilities and a higher material standard of living than most other places in the USSR. The strictly organized town centre accommodated all of the vital buildings and institutions (p. 166). The Cultural Palace at the northern end of the town square was outfitted with a sports hall, library, studios for music, dance and fine arts, and a theatre/cinema with a collection of several thousand feature films. Adjacent to the Cultural Palace was a large swimming hall, an indoor firing range, and the Jurij Gagarin Football Stadium. Further away were the mine bath (containing a sauna, showers, dressing rooms, and a relaxation department for higher ranking staff) and two generations of administrative blocks (the newest of which housed offices, laboratories, and a hidden department for the secret service[15]).

On the opposite side of the town square was a modern and well-equipped hospital, a school, a kindergarten, and the former Cultural Palace (this later housed the main kitchen and a mess hall for lower ranking staff). The town centre in Pyramiden also

included an ice hockey range, a radio station, lounges for workers ('Red Corners'), an outdoor playground, and a dance platform built to commemorate the 1980 Olympic Games in Moscow. In addition, there was the modern Hotel Tulip, which also housed a public bar, the post office, and the local museum.

Incorporated into the spatial layout of the town were more than 30 smaller and larger dwelling complexes and an elaborate apartment block for higher-level employees. The latter was located on the opposite side of the town square to the Cultural Palace and provided dining facilities, lounges for entertaining visitors, and a bar. Also located within the town centre were a brick factory, mechanical workshops, garages, a fire station (also containing a jail), a helicopter base, a greenhouse, a pigsty, a cow barn and hen house, a cemetery for cats set under a large ornamental metal sunflower, approximately 20 large storehouses for food and equipment, and one Norwegian traffic sign (warning 'Other danger' with subtitle 'Industrial area'). The town centre was surrounded by a harbour with facilities, storehouses, a power plant, a built-in railway for coal transport, a storage area for coal, fuel tanks, storehouse for explosives, the Blue Lagoon fresh water reservoir, a cemetery with 43 graves, and the 'bottle house' – a house for recreation made of 5308 empty bottles.

Due to the town's isolation during most of the year, what was needed in terms of personnel, advanced equipment, materials, and special food supplies was supplied in yearly shipments. Apart from this, Pyramiden was reliant on its own resources and thus constituted a remarkably self-sufficient town compared not only to the Norwegian settlements on Svalbard but also to most other Western urban settlements. The abundance of accessible coal created

an energy surplus that was used to feed and shape the town, which seems to operate and perform like one huge organism. All of its 'limbs', made up of buildings and infrastructure, were interconnected by a cunning system of pipes and cables for conveying hot water, cold water, sewerage, electricity, and telephone calls. Enclosed in wooden cages, they constituted a vein-like system that conspicuously traversed the surface of the town and also served as snow-free paths, supplementing and interconnecting the main streets made of large sections of reinforced concrete.

The coal-fuelled power plant supplied electrical power to the mine and the settlement; this was ostensibly the heart of the organism. Running the plant created a vast surplus of heated cooling water, which was piped back to the town, heating the saltwater swimming pool and all houses, and providing each apartment with plenty of hot water. The spill water from the power plant also warmed barns for cows, pigs and chickens, animals which produced milk, meat and eggs, as well as plenty of manure to fertilize soils in which cucumbers, parsley, tomatoes, green onions, chives, lettuce, peppers, and ornamental plants were grown in the greenhouse. In 1975, the agricultural yield was reported to be 35,000 kg of meat, 48,000 litres of milk, 110,000 eggs, and 5700 kg of vegetables.[16]

Turning coal into electricity also generated another useful by-product: huge amounts of mineral ash that were processed into concrete bricks at the brick factory. A major proportion of the buildings from the later development of the town were constructed of bricks produced in Pyramiden. The final waste from this cunning system of recycling was evidently very low, and resulted in a remarkably purified refuse dump. All food waste 'vanished' into pigs, cows and chickens. What little was left was covered

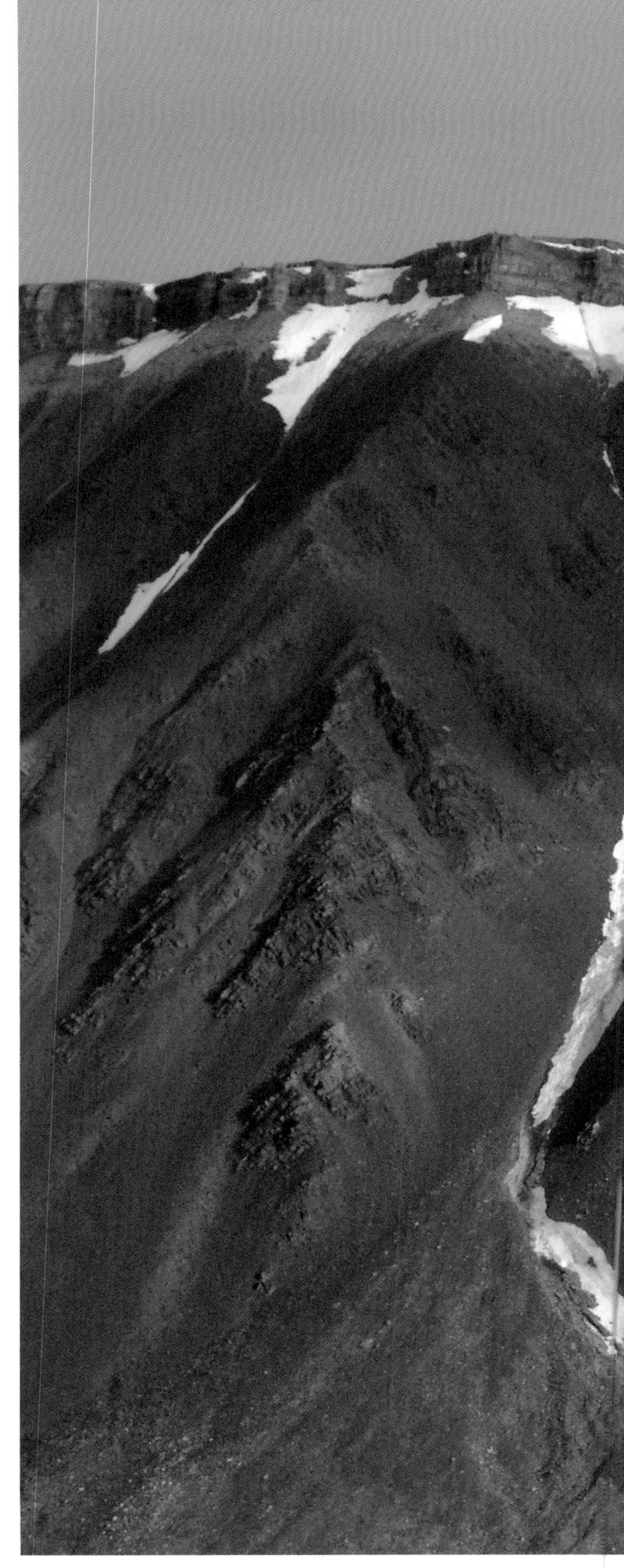

by the surplus of sandy ash from the power plant. As a result, the town dump comprised primarily a mound of sand with scattered cans, pieces of glass and metal, worn single boots, and other discarded personal belongings.

The self-supportive, recycling character of the settlement was also reflected in the abundance of locally produced tools and equipment, many bearing traces of careful maintenance, mending, and skilful artistic decoration. Such practices not related only to scarcity of supplies but also to a tradition of self-reliance and, of course, to low labour costs. A peculiar and thought-provoking expression of this material otherness is the 'Emergency Exit' signs which the Norwegian authorities demanded should be displayed at Hotel Tulip and in other buildings that occasionally might be visited by non-Russians. These luminous green signs, familiar to most of us – mass-produced and cheap – were here meticulously hand-made in compliance with the standardized global design: green squares were painted on the wall, letters cut out of white adhesive fabric were glued on leaving the pencil markings of the letters' contours still visible. There are many letters in 'Emergency Exit' – and in the Norwegian prescribed translation 'Nødutgang' – and there were many such signs at Hotel Tulip, the Cultural Palace and other public sites. Similar hand-made, mass-produced looking signs are probably extremely rare, even in the most socially extravagant of environments. Thus, their abundance in Pyramiden probably says as much about affluence as poverty.

The materiality of Pyramiden demonstrates an impressive functional totality. Dimensions, scales, power of equipment, facilities, tools, buildings, and staff were constantly up-graded to meet the mine's demands. The structure was dependent on and stuck with special functions and large-scale dimensions that increasingly imposed a strain on the net productivity of the enterprise. This was probably an important factor in Trust Arcticugol's calculations preceding the decision to abandon Pyramiden. The town was scaled for 1000 inhabitants, and in the mid-1990s the population gradually decreased to less than 500. The well-developed, but rigid structure was impressively functional at its peak, but lacked the flexibility needed to survive the new post-Soviet socio-political and economic conditions.

Building a none-place

All significant parts of Pyramiden were constructed prior to 1991, and its appearance makes manifest its material rootedness in the USSR era. However, despite these roots the town started out differently, being initially a small settlement attentive to the nature of the place. The buildings constructed in the late 1940s and 1950s were mainly concentrated on the dry and inactive alluvial fan at the base of the mountain and next to the mine.[17] Topography structured and guided the spatial outline of the early settlement; buildings were located and oriented largely in accordance with the will of the landscape. One example is the first administrative building, clearly the most prominent in Pyramiden's early post-war era. With its white-painted symmetrical façade it constitutes a very conspicuous architectural feature. It is hardly coincidental that the building is situated such that when approaching the main entrance the peak of Mt. Pyramiden is visible directly above it. Moreover, the tower-like appearance of the profiled gateway annex seems to mirror this dominant topographical feature.

The earliest apartment houses introduced a new and rigid linear outline to the settlement. If the aerial photo taken in 1960 is compared with the one from 1990 it is evident how constitutive this spatial pattern became for Pyramiden's later development. All of the apartment houses built during the 1960s were arranged symmetrically along a central axis that both ignored local topography and extended the town northwards into the active river delta area. When the town expanded extensively during the 1970s and 1980s all new buildings and infrastructure were spatially subjugated to this rigorous 'grammar'. The importance of the axis is also demonstrated by its canonization in 1977, when it was renamed 'The Street for 60 Year Jubilee of the Great October'. One outcome of this spatial regime was an increasingly more enclosed town square, where the views to the surrounding landscape became increasingly obscured. Standing in its centre, one would have been framed and encapsulated by the strict symmetry of the apartment blocks and the 'capital buildings'. The breathtaking view to the fjord and the glacier beyond was blocked by the apartment house for higher ranking staff. One would have beheld a built environment that could be literally anywhere in the USSR, including the repertoire of familiar material signatures.

Another crucial outcome of the linear town grammar was that Pyramiden could only expand by appropriating the very active riverbed to the north and west of the town. This risky expansion was made possible by several dikes and dams that successfully changed the course of the river and directed the water to flow outside the settlement. In addition to meltwater, the dynamics of permafrost also had to be tackled. As long as the ground remains frozen, all foundations will remain stable, but during summer, defrosted top-soils become water-clogged and unstable. However, Soviet engineers were highly experienced in Arctic construction and prevented defrosting by installing Freon-based freezing systems around large buildings in vulnerable positions. This overruling and taming of nature of course complied well with Soviet ideology and the communist wish-image of the world as man-made, where the forces of production liberated man from the constraints of nature. Nevertheless, the bottom line was that Pyramiden developed in a way that demanded constant management and intervention to protect it from the harsh Arctic conditions.

The Pyramiden that came into being during the 1960s, 1970s and 1980s, and as witnessed today by the surviving accumulated piles of material utterances, may be conceived as a peculiar Soviet version of the 'non-place'. Coined by Marc Augé to denote a new 'supermodern' *global* category of place, the non-place is characterized by its absence of cultural references, its denial of place and 'Heimat', typically exemplified by international hotels, airports, shopping malls, etc.[18] In short, it denotes a place that at the same time is everywhere and nowhere.

Soviet architecture and city planning were far from global, but still geographically extensive and materially effective. The particular signature of 'Sovietness' that was infused on sites such as Pyramiden and thousands of other smaller or larger towns still dotting the vast northern Eurasian landscape is easily recognized by its spatial rigidity, its monuments, concrete architecture, and iron installations.[19] The outcome was not only uniformity but also *dislocation*; a process of removal whereby local conditions, the culture/nature of place, became largely irrelevant to form and appearance. Pyramiden could have been a town situated anywhere within the Soviet realm.

Pyramiden started out as place; a small mining community that at least to some extent was sensitive to local conditions and the nature of place. However, during its heyday it was increasingly turned into a Soviet 'non-place', a site that denied or ignored its own location. The town square developed into a rigidly structured space of ideological performance containing the obligatory Lenin monument, the Pyramiden/Trust Arcticugol monument, and the free-standing poster boards announcing the worker of the month, official information and production rates.

Conspicuous, too, are the grey metal stands decorated with stars, and hammer and sickle, designed as bases to hold flags and banners; paraphernalia central to Soviet commemorative and political performances. Closer examination revealed that the ground contained numerous metal stands for flags and banners, as did the house facades, and suddenly it was apparent that the small crowns on every light post dotted around the town square were also designed to hold flags. During our fieldwork a total of 242 ready-made sockets for flags and banners was recorded in the square. In a matter of hours, the confined town square could have been transformed into an impressive arena for celebration and political re-enactment.

Contributing to the image of dislocation was also the fact that the town square was covered with imported, green grass, said to have been specially cultivated to cope with harsh tundra conditions. Today, much of this grass is still growing constituting a conspicuous contrast to the barren face of the surrounding Arctic ground. What could have been more powerful, more reconfirming of Soviet vitality and success than such a well-designed, man-made spot, with red flags and green grass in the midst of one of the world's most hostile environments?

The Soviet vision of Pyramiden, however, rested on an extensive regime of skilled management and surveillance that enabled it to perform *despite* its location. From the outside, stubborn Arctic rivers persistently reminded the town about the contrived nature of its very location. Dams were constantly under attack and had to be regularly repaired and defended. As long as the town was managed, water flows could be controlled, thawing soils could be kept frozen, and damage caused by mudslides, solifluction, wind, and snow could be repaired. The abandoned

Pyramiden, however, reveals its vulnerability and underlines the radical manipulation necessary to maintain its wished-for format.

The degree of vulnerability became apparent during our fieldwork. The river fed from glaciers to the north-east of the town may look innocent, but after a few days of sun, rain or warm winds there is a considerable increase in the amount of meltwater. The river water becomes saturated with sediments, more than it is capable of carrying after losing speed at the base of the steep valley. The river becomes clogged and forms braided streams that constantly change their course and, eventually, this glacial water succeeded in breaking through the primary dam that for decades had obscured the course of the river. Watching this 'return of the agency of the wild'[20] was an impressive, although somewhat frightening, experience. The untamed meltwater forced its way under the corner of the Culture Palace, around the Lenin statue, and along the streets of the central town. After it had eroded a course beneath the apartment block for higher ranking staff, it divided the playgrounds and proceeded to damage numerous storehouses.

On entering our home in the abandoned Hotel Tulip on the third evening

On entering our home in the abandoned Hotel Tulip on the third evening, I felt the light touch of something coming from above. A familiar smell soon revealed what: the oily regurgitation from one of our co-residing gulls. In my hair, on my jacket – this stinking fluid, fatty fish extract meant for the ever noisy baby birds nesting on the window ledges above. Inside the building, safe from the gulls, the smell lingered as an 'addendum' to the dry chalk-like odours emanating from crumbling paint and plaster that covered all things one touched. By now, this film of mixed matter from the decaying town was covering my skin, clothes and shoes. Soon, it would also be under my skin. By breathing and eating, I had let the dead city invade me. I felt a sudden urge to wash, for clean water, hot water, a shower in a heated room. The fact that Elin, Bjørnar and me were literally surrounded by several hundred bathrooms, sinks, toilets, showers, and 200 metres from a large swimming pool – without water, malfunctioning drainage and no heat – was just enhancing my feeling of being unclean. And water everywhere. Brown clay saturated meltwater from the adjacent glaciers – that had licked the surface of the left settlement. This water would make my hands wet – but not clean. It was time to make dinner.

I was assisting Bjørnar in cutting the canned ham for a macaroni casserole, trying to touch the food as little as possible, hoping that Bjørnar and Elin would do the same. I was thinking about our intriguing, strange and increasingly uneasy life in and among the buildings and material remains that were the object of our study. Using things left behind by faceless 'others', sitting in their chairs and eating at their tables, sleeping in their beds. Depending on things that to a large extent were deprived of their original functionality. Houses with walls and roofs, but colder inside than outdoors. Broken or open windows that let cold air flows live their own life. Nowhere to get warm. Marks from water running in the wrong places. Mattresses, not actually dirty – but not clean, not wet, but still not dry. That had developed a smell of their own, with no direct links to the previous, faceless sleeper. Still soft, however not any longer meant for sleeping – but left for a long and lonely disintegration. All things semi-usable, but also clearly marked by the early stages of decay that in time would transform them into archaeological structures and objects. Things that eventually would be cleaned by time, deprived of smells and substances that one could identify as belonging to the initial users.

This initial, ambivalent neither-nor stage is disturbing – and usually, in our contexts of domestic order, calls for action. We discard things, drag them to the dump, crush, burn or bury them – if not maintained, mended, dried, washed, repainted and reused. You do not like to see, nor smell, and certainly not like to live among things that decay. Odours, scents, stenches – you have to inhale them, drag them into your body to sense them. When you sense something 'bad' it is too late – it is already inside. Disgusting. But what is 'bad' and what is 'good'? If there was an objective scale for 'odour quality', would the stingy smell of fresh paint score higher than the mild and muffled scent from a slightly mouldy mattress? Is not the plusses and minuses of odours related to our notion of what is happening to things, that relate to their condition, their status in our daily intercourse with things? We are learnt to like the smell of things that are tended to, maintained, oiled, washed, painted, and renewed – including the

dangerous gases from paint and varnish. Vapours mediating function, value, care, and affection. Decay is equal to abandonment, neglect, throwing away, malfunction, contamination, death – all the pestering negatives we have learnt *not* to like. These things do not smell. They stink. Do not touch.

By now, I fully understood Elin's evening ritual of getting undressed on top of a plastic bag before popping into her sleeping bag in our suite in the fourth floor of the abandoned hotel. And our preference for having lunch on the flat, rain-washed roof of the hotel. Telling each other that this was a safe place from polar bears, and commenting on the magnificent view. But also the not-so-much-talked-about need to get away from disturbing odours and dusty, pasty substances. A place on the outside, that provided a necessary distance between us and the decaying town.

Later, dozing away with cold feet in our sleeping bags – there was a sharp sound penetrating the continuous cacophony of the gulls. Some metal sheet blowing in the wind – or? Elin asked me whether I remembered to bring the rifle upstairs. It was parked in the entrance hall on the ground floor, cartridges removed, in line with Norwegian regulations. We were about to fall asleep in the bull's-eye of the only smell of food (Bjørnar's macaroni cheese and ham casserole turned out to be very tasty) for miles and miles. We had observed fresh polar bear footprints by the power plant, and the possibility of a polar bear lurking around within the reach of the tempting odours was highly realistic. Was this the reason that Bjørnar had chosen his quarters in the far end of the corridor – as far as possible from our provisional kitchen? The drift towards sleep was punctuated by a notion that a polar bear already had entered the building, gently and silently pushing through the double main doors. I had to bring the rifle to my bedside. That meant leaving the warm sleeping bag, bare feet in wet boots, walking down the stairs, entering the first floor darkened by the wooden window shutters. The sound of my lonely footsteps echoing on the big staircase, doors everywhere, some open, some closed, and even more neither open nor closed. Wet boots were soon the least of my troubles. A feeling bordering on fear, disproportional to the certainty that there was absolutely nothing there. It is amazing how much dark emptiness may contain. Would I even dare to be here if nights were dark as they normally are?

Back in the sleeping bag, now with the loaded rifle in the corner by the window. The sleeping bag was not so snug anymore – it felt more like a body bag. Three flesh-filled cocoons – an unexpected bonus for a food seeking bear. Very unfamiliar to our modern minds, this raw fear of an animal that is (more than) willing and able to kill you. This was not a question of avoiding coming between mother and child, disturbing when eating, surprising, scaring, teasing, provoking. This animal was likely to observe, lurk, wait, and strike suddenly, when you least expected. When you had your mind on something else. An animal that would hunt you for the value of your flesh. In spite of all body fixations in our time, I had never thought of my body in this way – about its nutritional value for someone else. Was there a muffled sound of silent paws behind the noise from the gulls? Would I have time to wake up, find the small metal part at the top of the zipper, pull it down, get out of the sleeping bag, grab the rifle, charge and shoot? Would I hit my target? I was fantasying about adrenalin kicking grazing shots and a miraculous escape to the (hitherto) safe roof of the building. Then what …?

It was a mental effort to think about something else. An absurd thought that this once was a safe place for hundreds and hundreds of people, living normal lives in a functional town. That I once experienced Pyramiden with its proud people. I was thinking about how I always liked to be here. As Cultural Heritage Officer in the Norwegian Governor's staff I attended several arrangements and meetings, especially during the hectic months when mine and settlement was closed down in 1998. I had actually stayed in this very building when it was a proper hotel, when somebody still painted the charming metal tulip by the entrance in bright red, yellow and green. All things so normal that I hardly noticed any of them. White sheets, rooms tended, clean, dusted, heated, dry, safe. Clean water running from all of the taps, hot water in the one marked with red, toilets that flushed when one pulled the string, showers that drained, electricity from all plugs, light from all lamps when you turned the switch, mirrors above all sinks. Sounds of footsteps, passing trucks, somebody hammering on something, doors that were opened and closed, voices … I remembered how Russian always sounded angry to me, especially when reverberating in a long corridor. The smell of food, green plants, detergent, the very masculine Russian aftershave and likewise feminine women's perfume. White, fresh bread

that produced slices bigger than your plate. At that time, it was absurd to imagine the town without its people.

My first visit – it was the celebration of the Russian Christmas in January 1997, and an arrangement in Pyramiden's Cultural Palace. Arriving in the middle of the three-month long winter night, on-board the Governor's helicopter, along with other Norwegian representatives and officials, the Governor's interpreter, the Longyearbyen priest, and a children's gospel choir. The flight from Longyearbyen was twenty peaceful minutes in total darkness before seeing the chains of road lights, all converging on the town's centre. On the snow-covered heliport, footsteps made muffled, squeaky sounds. I tried to breathe slowly through the mouth to prevent the unpleasant ice gluing in my nostrils. A warm bus with curtains, a powerful engine making the whole vehicle vibrate. Orderly ploughed roads, sharp, clean and defined. Warm lights from most windows, a multitude of curtains producing a mosaic of colours. The pristine covering of snow around the houses – like a child's drawing – simple, tidy and inviting. People everywhere, producing puffs of frosted mist as they busily moved from this to that along neatly hand-shovelled short cuts or the network of pipeline cage walkways. They had things to do, bring and fetch. Work, appointments, leisure activities, rehearsals, and social calls. Women with bright red lipstick, fur hats and coats – an urban touch very different from the Longyearbyen Arctic wind-stopper Gore-Tex dress code. The giant handmade outdoor thermometer at the low end of the town square: a row of light bulbs showing −16°C. The wall of heat hitting my red-cheeked face on entering the crowded Cultural Palace.

The arrangement itself started with a chain of introductions, each ending with passing the word to a higher ranking representative, the highest one undertaking the official opening. Russian and Ukraine folk dances and traditional songs. Bright colourful costumes, precise accompaniment from acoustic instruments, all movements thoroughly rehearsed and choreographed down to the most minute details. It was an effort to remember that all this was produced by a community specially designed to extract coal from the depths of Mt. Pyramiden. A man was video recording the whole event with a camera bigger than we had seen in many years. In between the artistic sequences, speeches and greetings, friendly, polite exaggerations about things that nobody would hesitate to approve of – such as my own words spoken and translated into Russian from the stage (I still have the manuscript):

> Dear Ukrainian and Russian friends
> *(Russian translation)*
> On behalf of the Governor I would like to thank for the good cooperation between Russian, Ukrainian and Norwegian Svalbardians in 1996. We give thanks for the year that has passed, and wish all a happy new year in 1997.
> *(Russian translation)*
> We Norwegians started our Christmas celebration already in November, and we are now happy to have normal work days. By now, we have replanted our Christmas trees along the ski track in Longyearbyen, and are enjoying our little wood as we exercise to get rid of all the Christmas food.
> *(Russian translation, followed by scattered, polite laughter)*
> [...]

The translations gave me plenty of time to check my next phrase, and to study the response from the audience, calming my nerves. I was impressed with how firm and confident my speech sounded in Russian

(highly undeserved, as the Governor's experienced interpreter added a punch that my original performance lacked). And then, the jewel of the evening, the Longyearbyen children's choir – filling the big auditorium with Norwegian Christmas carols, perfectly balanced between cheerfulness and melancholy. This was the undisputed highlight for many, as only a few high ranking employees were allowed to bring their children to Pyramiden at that time. Handkerchiefs drawn from pockets indicated that the perfect balance was disrupted towards melancholy that provoked profound longings for little ones not seen in almost two years.

Afterwards, an impressive buffet was offered to the Norwegian guests. The children were well entertained in a separate room, giving peace a chance for the adults. A tempting table, tasty and beautifully prepared food – freshly baked bread, varieties of sliced meat, cooled deep-fried fish, pickled carrots, beets and cabbage, sausages, chicken legs fried in breadcrumbs, locally picked and dried mushrooms baked in sour cream. It was hard to comprehend that this community received supplies only once per year. Crispy green cucumbers and parsley ('local produce ... from our greenhouse'), mineral water, beer, wine – and the vodka. In his welcoming speech, the director underlined that drinking was bad, but toasting was culture – and we made a toast. The second toast, in line with unspoken etiquette, was to the Norwegian party, and the Longyearbyen priest thanked our hosts for their invitation, and complimented them on the arrangement we just had enjoyed. The language barrier and subsequent translations added a pleasant solemn taint to the session – no small talk, all things said in plenum. After a few mouthfuls of a new set of Russian and Ukrainian snacks, all vodka glasses were once again filled to the rim in readiness for the third toast:

Dear friends – as you know, it is our tradition that the third toast is in tribute to the women. We Russians are very proud and fond of our women. We love, respect and adore them. In fact, we think that women are man's best friend [a friendly provocation directed towards the Norwegian self-righteous focus on gender equality, and the female Governor of Svalbard at the time]. I ask you to join me in this toast – and let I remind you that as many drops that remain in our glasses, our women will weep for us.

Strange, this ability of vodka to enable friendliness and compliments flow effortlessly. The only one in the party that had reason to worry was our interpreter, forced to join in all of the toasts, but given no time to eat. He really needed his long experience. At the end of the meal we had agreed to meet in Pyramiden as soon as the snow was gone, to climb together to the summit of Mt. Pyramiden, where, under the Russian Red Star Monument, we would make a toast to pay tribute to the friendship between our two nations for all future. How unthinkable, this afternoon, that Pyramiden would be abandoned in less than two years.

The town without people was even harder to comprehend after the decision to close down Pyramiden was clearly expressed in the autumn of 1997. On the many official visits during deconstruction I was able to observe the gradual decrease in the staff and a subsequent decrease in activities, confined to fewer and fewer buildings. I recall a happy afternoon in the excellent salt water swimming pool, but also the disturbing twinge in constantly remembering that all things around me were on the verge of being surrendered to varied but merciless processes of decay, a long and degrading one way journey to the Eternal dust. All things sensed would be gone – the fresh, humid, almost tropical smell of the heated salt water, the reverberations from people enjoying the swimming pool, the cool sea-green tiles and charming wall mosaics that would fall down. On returning to the Hotel Tulip, I was asked to photograph a group of friends by the Pyramiden monument in the town square. It was their last evening here. They were cheerful, but I sensed sadness too. Maybe my own?

The town, soon to be emptied of its human members, was to have an unexpected effect: a change of focus, an increasingly awareness of all material members of the community. In being left alone, no longer in their intended function, places, rooms and things became increasingly tangible. Not only as remnants and refuse from human actions, but also as active 'members' of the community, making the living settlement possible, regulating movements, actions and behaviour. Between them, they revealed layers of unintended evidence of how the community was managed, its political superstructure, peoples' abilities, and traditions.

On my last visit to the living Pyramiden, after the coal-fuelled power plant was closed, it was like seeing a dying friend in a respirator. The official meeting between Trust Arcticugol and the Norwegian authorities was like deciding what to do with the dead body, how to honour the memory and minimize the unwanted side-effects of decay. Norwegian environmental authorities were heavily represented at the meeting: apart from the Governor's staff there were delegates from the Directorate for Nature Management, Directorate for Cultural Heritage and, not the least, the Norwegian Pollution Control Authority. Soviet industrial enterprises are seldom prized for their environmental profile. Now, the Russian state company was

about to leave after fifty years of 'dirty' activities in the heart of Svalbard's pristine environment. The session started with mutual orientations and polite exchanges of visions for Pyramiden as an abandoned settlement. Later, a guided tour of different parts of the settlement, showing the closed 'capital buildings', sites of removed buildings and various dumps and deposal sites. There was a growing suspicion that places with low misery factor were overrepresented. We wanted to see the city dump.

The director's red minibus transported us to the big heap on the far side of the valley. It seemed like sand, as clean as that in a playground sandbox – not like any city dump any of us (including the representative from the pollution control) had seen before. Of course, we knew that this was not sand, but ash – the mineral component in the coals that had fuelled Pyramiden's power plant. Apart from some minor fragments of this and that, there was hardly anything of what we expected to find at the main fill site of a long lasting industrial enterprise. A hole with still smouldering refuse (broken glass bottles, jars, cans, plastic wrappings, crushed wooden pallets, a few bones, a pig's skull) was observed, ready to be covered by ash. With reference to our own accelerating dumps in Norway – even in Longyearbyen – something was wrong: there had to be more. We would not be fooled. The representative from the Norwegian Pollution Control Authority would not leave without penetrating the dark secret of the Pyramiden dump. I recall a conversation that more or less went along the following lines:

> Pollution Control representative (PC): So ... this is the main dump in Pyramiden?

Trust Arcticugol representative (TA): Yes.

PC: But where is the organic refuse, domestic refuse ... the food remains?

TA: Most of that does not end up here. This is not refuse – it is a resource. We feed this to our pigs and chickens. In fact, our food remains are a large part of what we feed to the animals. We also have a deposit for animal manure by the pig house that we use in our greenhouse and in the many flower pots in the settlement.

PC: Well ... how about disposal of chemical waste ... is this also going here?

TA: Chemical waste – what do you mean?

PC: ...like ... paint, for instance?

TA: But why should we throw away paint? We hardly have enough paint for maintaining our buildings. We certainly do not throw away paint.

PC: ...and equipment with components that may cause environmental damage, such as industrial batteries?

TA: Ah, accumulators. They are very expensive – and contain valuable components. We ship those back to the mainland for recirculation. The ships that bring our yearly supplies have a lot of free space on their return, for accumulators and other valuable components that may be reused.

PC: But what about smaller, non-chargeable batteries ... for transistor radios, cassette players and other battery operated equipment?

TA: There are not a lot of these – some have private cassette players, but for the most part they are plugged to the mains. Batteries are expensive and not easily accessible here. But yes, I have to admit – the ones we have had have ended up here ...

I recall overlooking the approximately 100 metre wide and 10 metre high pile of seemingly clean ash, envisioning the positions of … perhaps 138 batteries. Or 1380. Regardless, Trust Arcticugol would probably not score very highly among environmentalist-profiled industrial enterprises. Without doubt, there is an abundance of unhealthy chemicals in and around the town, barrels that threaten to leak – things sucked into the ground – bad things seeping out of the large ash deposits. In recent years, measurements in Billefjorden have demonstrated alarming levels of PCBs (polychlorinated biphenyls)[21]. The Trust's representative may have known more than he liked to talk about – without receiving a direct question. But this is not the point here. The city dump revealed a different relation to things. Compared to dumps in Norwegian and Western communities, the Pyramiden dump contained considerably less refuse, and it also differed in its composition. Obviously, this community reused and recirculated more of their things. Not for environmental reasons – but simply because it paid to do so. The combination of considerably lower labour costs and fewer resources to buy new things implies that maintenance, reuse, repairing, and dismantling pay off. One may therefore wonder to what extent Western well-organized and highly fronted refuse handling is a necessity resulting from our use-and-discard relation to things. Are our elaborate and highly fronted waste-handling strategies a not-too-much-talked-about must to maintain this high turnover consumption pattern that is so vital to Western economy, or a pure and altruistic concern for the planet's health?

This is getting close to our reasons for enduring the hardships of the abandoned settlement – why Bjørnar, Elin and me visited the remaining material inhabitants of Pyramiden. They were integral parts of a community contemporary with our own, and we recognized most functions, contexts, combinations, and placements. Still, as demonstrated by the Pyramiden dump, other principles ruled this community, and things are different. These discrepancies bring things to your attention. You may not understand all you see – but at least you notice. And here – in the post-human settlement, we were given unrestricted and uncensored access to all material 'informants' – an access that Pyramiden's human residents probably would have restrained or resisted if they were still here. It was an option for holistic (but not all encompassing) perspectives on this community ... without the worries of intruding on privacy. The material 'residents' do not seem to mind, I guess they have nothing to hide. Things may not always be what they seem to be, but they are always true. Their relations to their human co-players may be intricate and subtle, but they are always true, interlocked in a web of rationales to other things, to humans and human actions, society, beliefs, rules, and preferences. This simple fact is often drowned in the awareness of 'source criticism' – our meticulous concern for all possible effects of non-human agencies that must be subtracted and annulled before even thinking about interpreting material remains. We seem to forget the most basic – things may corrode, break, fall down, be moved, buried, or burnt. But they have no interest in pretending, deceiving or concealing. They will serve ... and rule whoever comes by.

PLAYBOY
314

Making a home

Workers that came to live and work in Pyramiden were normally hired on two-year contracts. The contract period was followed either by a free visit home or a long summer holiday for those who signed on or were offered a new contract. During the contract period, the Trust supplied food, housing and working clothes for employees – wages were deposited in their bank accounts for the benefit of their closest family members. In the most active years, Pyramiden housed a family society, with couples and children, but for the most part the inhabitants (especially the miners) who lived far away from their families in Donbass, Donetsk and Tula were single. How did such people cope in Pyramiden? How did they relate to the dominant materiality and the 'bio-politics' it facilitated in terms of appropriate public behaviour? To what extent was Pyramiden – also during its phase of human occupancy – a contested place?

The public spaces in Pyramiden were conspicuously free of any manifestations of opposition and resistance. Despite the seven years of post-Soviet existence, the town's facade revealed little other than a wealth of Soviet 'affordances':[22] monuments, kitsch, and posters. In the administration building the collected works of Lenin were still on the shelf in the director's office. Signs of 'otherness', deviant behaviour, were absent. Perhaps what was most amazing was the lack of graffiti, even from the post-Soviet years. The built-in hallway that ran for several hundred metres from the mine bath to the cable car, and which provided a sheltered walkway for hundreds of shift work miners every day, literally begged for graffiti – yet none was found.

The workers' apartments, however, provided a very different impression. We had the opportunity to survey apartment block 38 located on 'The Street for the 60 Years Anniversary of the Great October'. The block contained 72 small flats, and apart from those on the first floor they each consisted of two rooms: a small front room with an annexed bathroom and a larger combined sleeping and living room. These flats were designed for singles or couples without children.

The corridors and stairways of the block mirrored the public spaces: standardized, orderly and no graffiti to be spotted. However, as we approached the doors, differences started to emerge: glued-on labels, signs, paint, handcrafted door numbers, and, in one case, an elaborated wood-panelled door. Inside each apartment we found an amazing wealth of differences and decorative inventiveness.

Despite their spatial uniformity, no two of these flats look alike. Creative use of wallpaper, floor coverings and paint made each apartment unique and individual. Flowers, potted plants, self-produced furniture, and inventive bookshelves added to these idiosyncrasies, while the wall embellishments made manifest the sublime bricolage of using whatever was at hand for decoration and difference. From cigarette boxes, beer bottle labels and advertisements to pin-ups and air cargo package tape; in short, most materials seemed feasible. All this made the interiors of the apartments stand out in astonishing contrast to the purity and disciplined utterances dominating the public spaces. Inevitably, elements of the official discourse and body-politics were intruding even here: dumbbells are found everywhere, as are training bars above the doors.

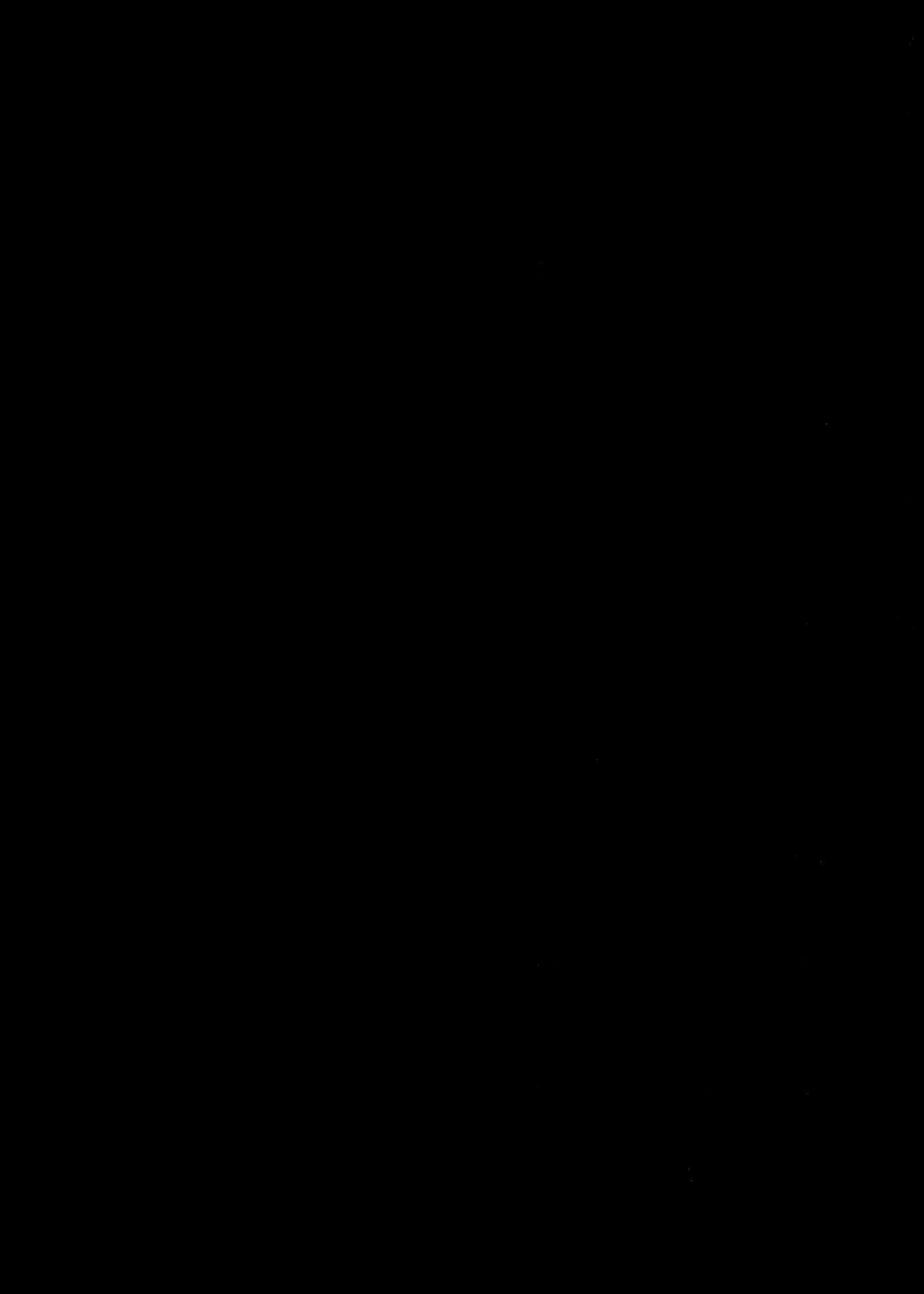

There are probably varied and ambiguous reasons for the expressiveness and otherness of the interiors. For one thing, people may have symbolically and materially appropriated their flats to fight alienation, conformity and boredom.[23] By adding their personal touch in the form of wallpaper, paint, flowers, and home-made furniture and decorations, the occupants turned the uniform flats into potentially inalienable and creative possessions. In other words, they made them into *homes*, into something that helped secure their personal identity and well-being. The workers that arrived in Pyramiden from places such as Donetsk, Donbass and Tula clearly found themselves very far away from home, and they faced a potential threat of alienation that should not be disregarded. Considering Pyramiden's dominant materiality, there was obviously a need to carve out something closer and more personal.

However, the variety and specificity of the domestic materials also calls for other explanations. For instance, we often forget that such manifestations also are expressions of *skills*. In a society where so much relied on what was home-made,[24] there was probably pride in being a skilled door maker, a talented glass painter, an adept shelf builder, a resourceful bottle-house builder, or a creative cigarette box decorator. In short, a silent competition of being the best bricoleur. In Pyramiden possessions and relative social status were, to some extent at least, more dependent on skills than money. This, of course, also involved the skill of networking and bartering, obtaining access to the 'prestige-goods economy' where the desired raw materials – advertisements, magazines, cigarette boxes, paint, panels, labels – were circulating. Some of these goods were accessible only through the Norwegian society in Longyearbyen, while others

came from local workshops and storehouses and were embedded in their own networks of restricted exchange.

The actual materials mobilized to decorate these interiors suggest other agendas as well. There are subtle – and some more overt – political messages in Pyramiden's home decoration. In one apartment, a portrait of Karl Marx was found properly framed on the wall, while on a sofa bed in another apartment laid a poster of an over-decorated Leonid Bresjnev being awarded even more cut-out paper medals, and with a bullet-like glue mark in his forehead. Furthermore, the 'raw materials' used for decoration were not chosen arbitrarily. As already noted, what they displayed and offered is very much what public spaces and Soviet ideology are *not* about. The imagery of the apartments represents a kind of inversed expression of official living and ideology. Wall decorations were dominated by glamour pictures and advertisements for capitalist consumer goods.

There were, of course, also the compulsory pin-ups, but there were also images not complying with the dominant heterosexual regulatory ideal. There was variety and some enigmas in these expressions, such as the pictures of polite bureau members found on the *inside* of a wardrobe door. However, it may have been significant for the tropes of domestic performance in post-Soviet Russia: pin-ups[25] were moved to front stage and the images of the old regime were moved backstage.

It is also interesting to note the contrast between the miners' homes in block 38 and the flats occupied by the officials. The latter were located in the big apartment block constituting the southern confines of the town square. This block was set crosswise to all other apartment houses, sharing orientation in the town centre proper only

129

with the Cultural Palace located at the opposite side of the square. The semiotics of material differences was also otherwise crude. The higher ranking staff lived in bigger apartments with the comforts of separate bedrooms and proper kitchens. Even the compulsory bars for training the occupants' biceps were bigger and more elaborate. The signs of distinction also included such extravagances as wallpaper on the ceiling, later increasingly victimized by the joined forces of crumbling glue and gravity. More important, however, none of the officials' flats contained the cheap advertised images of Western consumer goods found decorating the walls in the worker's flats. This may have been due to the fact that the officials could afford to buy the 'real thing', or, more likely, because such acts were politically incorrect. The surveillance and control over these spaces was probably far stricter than that of the miners' flats, and a wrong touch in your home decoration could have proved fatal for further career moves. Hence, no pin-ups or pictures of film stars, no glossy Seiko advertisements, and no artwork based on the cut-out boxes containing Brooklyn cigarettes. The walls seemed mainly to have been kept clean and if any decoration remained it consisted of huge wallpaper-posters depicting secure images of pristine Russian nature.

A ruin in the true sense of the word

The impression of Pyramiden as a fossilized town preserved for eternity is, of course, far from true. Despite its remarkable wealth of materials in place, Pyramiden is also aging. At the time of our visit Arctic rivers were already reclaiming their courses, gulls were annexing former human spaces, wallpaper, plaster and paint were withering and crumbling, and solid materials were disintegrating. In other words, the town we encountered was also a site of decaying modern debris, a ghostly display of things made redundant – in short, a ruin in the making.

Conceptualizing Pyramiden as a ruin, however, is far from as trivial as it sounds. In modern thinking the term is inflicted with its own values and effective-historical traditions, most notably expressed in romantic art and heritage discourses. It is also seriously affected by the oppositional regime that urges us to separate things functional and/or aesthetically pleasing from *waste* – all rubbish conceived of as litter and pollution and supposed to be eradicated by increasingly more effective systems of disposal and recycling.[26]

Heritage practices may at first be seen to be mediating this opposition, reflecting a care for and attentiveness to the useless and stranded. Heritage, however, contains its own regimes of cultural valuing and 'othering'. As Tim Edensor has argued, the common 'heritage ruin' is often a staged, neat and picturesque site that provides visitors with a disciplined and purified space. The presented rock carvings, church ruins and megaliths are all well maintained and neat; extraneous materials – plants, fauna, debris, modern materials – all 'matters out of place', are to be expunged.[27] Withering

and further decay is attempted staved off through restoration and preservation. Arresting decay, preventing 'death' has, of course, always been the imperative of modern museums and heritage management: preserve before it is too late!

Pyramiden is a ruin, but it hardly fits into the common tropes of heritage. The two posters placed next to each other at the door of the Hotel Tulip by the Norwegian heritage authorities may signify some of this ambiguity. One wished 'Welcome' to the town, while the other stated 'Private property' and 'No entrance'. One may well sympathize with the perplexity. What are we to do with this rusting and crumbling ghost town situated in the midst of pristine Arctic nature – Europe's last authentic wilderness? Moreover, in the dominant conception ruins are *old*, they have an 'age value' which is imperative to their legal and cultural-historical appreciation. Judged by this criterion, Pyramiden become ambiguous, even anachronistic. In its hybrid or uncanny state it becomes a kind of antonym of the modern and also blurs established cultural categories of purity and dirt; in short, it becomes matter out of place – and out of time. Add to this that Pyramiden is Soviet, industrial and utterly northern, and one knows that it hardly aspires to the ranks of the World Heritage List. Pyramiden is anti-heritage.[28]

Why bother then? Why is Pyramiden important? Apart from being a unique source to the lives of those Russian and Ukrainians who lived there, the unpolished and undisciplined ruin may teach us some alternative thing lessons. It confronts our customized habit of dealing with things as goods, as tamed domesticated possessions. Pyramiden takes us beyond consumption; the material is allowed to be itself. Things appear neither as frames nor backgrounds, but at centre stage. The *forms* of things are foregrounded:[29] their textures, their smells, their utter silence. In Pyramiden the *being* of things is hard to ignore. It is present, pestering – providing the visitors with an affluence of uncanny affordances.

The fact that Pyramiden is decaying may be seen as diminishing its potential thing lesson. Decay is usually understood only in a negative way. Things are degraded and humiliated through material alteration, while the information, knowledge and memories embedded in them become lost in the process.[30] Less attention is paid to the fact that things actually may release some of their meaning or generate a different kind of knowledge precisely through processes of decay and ruination.[31] Experiencing a working, populated city or a complete building may not reveal much about the way they actually work, the diversity of materials and technologies that are mobilized to construct and operate them. These 'comrades' tend to be cunningly hidden or disguised by design and smooth architectural form.

Ruination disturbs the taken-for-granted; in the destruction process new layers of meaning are revealed, meanings that may only be possible to grasp at second hand when no longer immersed in their withdrawn and useful reality. Ruination can thus be seen also as a *recovery* of memory;[32] a 'slow-motion archaeology' that exposes the formerly hidden and black-boxed. As expressed by Tim Edensor,

> Ruination produces a defamiliarized landscape in which the formerly hidden emerges; the tricks that make a building a coherent ensemble are revealed, exposing the magic of construction. The internal organs, pipes, veins, wiring and tubes – the guts of a building – spill out … The key points of tension become visible, and the skeleton – the infrastructure on which all else hangs – the

pillars, keystones, support walls and beams stand while less sturdy materials – the clothing or flesh of the building - peels off.[33]

Further, in proper ruins, culture and nature find their way back together, returning from their purified diasporas as entities set apart. Nature intrudes and mingles, confronting our taken for granted conceptions of the world as orderly and divided with matters all in place. Experiencing this process of re-hybridization adds another dimension to the portfolio of otherness that Pyramiden has to offer. Our conceptual boundaries are instantly alarmed by window ledges packed with nesting and cackling birds pressing themselves against the glass: by feathers, faeces and the rotting flesh of dead gulls on the floor in a former official's apartment – in short, by birds out of place[34].

Persistent memories

In experiencing Pyramiden, one finds truth in Walter Benjamin's saying that a ruin often speaks more honestly, more revealingly, than a complete building. The ruin proper, moreover, gives face also to that which is untimely and unsuccessful, bringing forth what are fragmented and lost in conventional history. It is 'physically charged with history', not as a treasure for eternity, but rather in the condition of 'petrified unrest'.[35]

To Benjamin, history decomposes into images, into historical objects, not into narratives. Historical objects are blasted out of continuous history, and become 'dialectical' images made actual in the present.[36] And what could be more blasted out of continuous history than Pyramiden, a residue from a stranded political and economic venture? Yet precisely by being redundant and discarded, this leftover town also reveals the gaps in the construction of history as progress, as a continuous narrative. Ruins such as Pyramiden, despite being utterly marginal in the current political economy of the past, therefore have their own historical mission: they rescue a forgotten past, not as heritage, at least not in any ordinary sense, but as a kind of *involuntary memory* that illuminates what conventional cultural history has left behind. They bring forth the abject memories that this history has displaced.

In Benjamin's conception of the dialectical image the past comes together with the present —not in harmony or as a fusion of horizons, but in a tension-filled constellation, a power-field illuminating *both* a fore- and an after-history.[37] This is precisely what Pyramiden does. In its ruined state it stubbornly carries the means to trigger the involuntary memories of its untimely past. As a peculiar site of

remembrance, Pyramiden presents us with the prehistoric 'wish images' of socialism, the already outdated dreams inscribed into its monuments, kitsch and town planning. These 'residues of a dream world'[38] are present everywhere in the face of this ruin, from the grand spatial boldness of the very site to the pencil marks left in the meticulously handmade emergency exit signs. They are visible in the street sign persistently telling us its pride as 'The Street for the 60 Year anniversary of the Great October'. They are also present in the slogan written in wooden beams on Pyramiden's steep mountain slope. The sickle and the hammer of this literally very moving text have already succumbed to gravity, but still visible are the two last words of 'Mir Miru Mir' – Peace on Earth, under which was later added 'Tula', the name of one of the Russian home towns. Although mostly confined to its back stages, Pyramiden also presents us with the 'too early' *ur*-phenomena of a utopian consumer society, of which the glossy images were mostly what were left for the miners to consume. In short, Pyramiden reveals to us the 'ur-forms' of socialism and capitalist consumerism, both of which are dormant in its after-history, in its fate as rubble in the present.

Due to its very chronometric proximity Pyramiden may appear as a contemporary site. Judged by the current material-political conditions, however, Pyramiden confronts us like a relic from a distant and discarded past. Already before the time of abandonment Pyramiden was a historically postponed site, surviving as a conspicuous Soviet town in a post-Soviet era. The crucial point is nevertheless that it is still present, most of it at least, and thus makes a difference. Pyramiden upholds a forgotten past, and makes it present and tangible. Furthermore, in the physiognomy it provides to history, Pyramiden continues to seize the ambiguous dialectics between the 'too early' and the 'already gone', so constitutive of its own historical fate.

Pyramiden in Svalbard

Svalbard is part of the Kingdom of Norway, and it is Norway that ratifies and enforces the legislation that applies to the archipelago. 'The Svalbard Treaty' (1920) provides for Norwegian sovereignty, and also confirms certain rights for the other signatories. 39 A total of 39 countries are registered as parties to the Svalbard Treaty. Citizens and companies from all treaty nations enjoy the same right of access to and residence in Svalbard. The rights to fish, hunt or undertake any kind of maritime, industrial, mining or trade activity are granted on equal terms. All activity is subject to the legislation adopted by Norwegian authorities, but the Treaty does not allow for any preferential treatment among the signature states. This is the formal legal basis for the enterprise of the Russian state-owned mining company Trust Arcticugol at Svalbard.

The Governor of Svalbard is the Norwegian Government's supreme representative in the archipelago. 'The Svalbard Environmental Protection Act'[40] is a collection of updated environmental legislations for Svalbard. It deals with area protection, species management (flora and fauna), cultural monuments and artefacts, land use plans, pollution, waste disposal, and traffic. The Act's 'Chapter V. The cultural heritage' states that

> Structures and sites and movable historical objects in Svalbard shall be protected and safeguarded as a part of Svalbard's cultural heritage and identity and as an element of a coherent system of environmental management. (§ 38)

The following are protected (§ 39):

- structures and sites dating from 1945 or earlier;
- movable historical objects dating from 1945 or earlier that come to light by chance or through investigations, excavation or in any other way;
- elements of the cultural heritage dating from after 1945 that are of particular historical or cultural value and that are protected by a decision of the directorate.

With minor exceptions, the present Pyramiden is a post-war construction and not protected by the Act. This does not mean that the town's value as a historical document is neglected by the cultural heritage authorities. In fact, the Governor of Svalbard encouraged and cooperated with Trust Arcticugol in securing the buildings when the settlement was abandoned in 1998. The Governor is also assisting the Trust in protecting the town from unruly visitors after 1998.

All constructions and other manmade remains in the settlement are the legal property of Trust Arcticugol, and any visitors to Pyramiden must respect this: artefacts and structures must not be damaged, moved or removed, and visits to buildings must be authorized by the Trust.

1960

1990

1 Road to water reservoir Blue Lagoon
2 Dike preventing river erosion
3 Storage house for feature films
4 Covered firing range
5 Cultural Palace (sports hall, theatre/cinema, library, studios)
6 The Jurij Gagarin Football Stadium
7 Lenin monument
8 Swimming hall
9 Small apartment blocks
10 Hospital
11 Mine bath
12 Covered walkway from mine bath to cable car/mine
13 Training facility for fire in mine
14 Brick factory
15 Hammer and sickle / 'Miru Mir'
16 Water works
17 Cable car to mine and covered conveyor belt for coal transport
18 Storehouse
19 Large apartment complexes
20 'The Street for 60 Year Jubilee of the Great October'
21 Pyramiden monument in town square
22 Administrative building
23 Initial administrative building
24 Covered railway for coal transport
25 Outdoor (summer) pigsty
26 Hen house
27 Storage area for bricks
28 Pig house
29 Cow barn
30 Animal manure deposit for greenhouse, flower pots
31 Greenhouse
32 Main kitchen and mess hall, initial Cultural Palace
33 Block 38, large dwelling complex under construction
35 Hotel Tulip, bar, post office, museum
36 Dwelling complex for higher ranking staff, representative lounges
37 Sunflower monument, communal grave for cats
38 Playground
39 Dance platform
40 Mechanical workshops
41 Road to ash and garbage deposits
42 Storage for barrels
43 Heliport
44 Garage, workshop for vehicles
45 Storehouses
46 Pipeline cage walkway, part of 'The Street for 60 Year Jubilee of the Great October'
47 Ice hockey range
48 Fire station, jail
49 Radio station
50 Oil tanks
51 Road to harbour, coal deposit, power plant
52 The Finish houses, oldest dwelling houses
53 Storage area for equipment, building materials

Pyramiden biography

On 29 July 1920, Soviet Union authorities passed an act aimed at supplying northern Russia with coal from Svalbard. The act was signed by the chairman for the Counsel of People's Commissionaires, V. Uljanov (Lenin). This was probably an important basis for the union's decision to acknowledge (1924) and sign (1935) the Svalbard Treaty, thus obtaining rights to coalfields on Svalbard.[41]

The Russian state-owned Trust Arcticugol was established in 1931. The rights to the coalfields at Barentsburg and Pyramiden were acquired in 1932.[42] The earliest development of the mine started in July 1939 – storehouses, fuel depots, a dwelling complex, and baths were erected at the base of Mt. Pyramiden and on the shores of Mimer Bay. The development of the settlement continued in 1940, in parallel with geological surveys and the construction of ventilation shafts for planned mines. The first overwintering was in 1940–1941.[43]

After the Nazi regime's attack on the Soviet Union on 22 June 1941, preparations were made to evacuate Pyramiden. On 25 August, a large convoy of British marine vessels arrived in Isfjorden, and orders were given for immediate evacuation of all Norwegian and Russian settlers. The 99 inhabitants in Pyramiden were given three hours to gather their personal belongings – no other cargo was allowed on-board. To prevent the benefits of Pyramiden falling into German hands (Norway was already occupied), the British forces set fire to deposits of coal and fuel, diesel generators, trucks, and tractors, destroyed machines and equipment, and blew up all explosives. Most buildings were left unharmed.[44] After the war, the Soviet Union suffered from severe energy shortages – the coal mines in the vicinity of Moscow and Donbass had been severely damaged. On 29 August 1946, the Soviet regime decided that Trust Arcticugol was to develop the Svalbard coal mines. Two mines would be established in Pyramiden, each with a yearly capacity of 300,000 tons of coal. The 29 August 1946 is recognized as the formal foundation day of Pyramiden.[45]

The parallel development of three mining complexes, Barentsburg, Grumant and Pyramiden, was a gigantic operation. In Pyramiden, the first workers arrived on 16 November 1946. Initially, Pyramiden was the Soviet capital on Svalbard and the first Consulate was established here. Liv Balstad, wife of the Governor of Svalbard, has described a visit to Pyramiden in 1947.[46] Although the settlement was characterized as consisting of 'dull and dreary barracks', they were welcomed by Consul Fetchin in his 'marine blue uniform with sturdy epaulettes and rattling sabre' in his newly erected Consulate. The following reception turned out to be a show of hospitality, entertainment and food that the visitors were little prepared for. Balstad reveals how she mistook a small afternoon snack for a full dinner, and afterwards, while watching fairy tale and football films, was 'deadened as a constrictor that had swallowed a donkey', and yet eager waitresses were still filling the slightest free space on her plate with new kinds of snacks, cakes and sweets.

The development of the settlement and mine accelerated during the next few years. In 1950, Trust Arcticugol shipped 1712 workers and c.28,000 tons of equipment and building materials to Svalbard.[47] In his book 'The Coal Mines at Spitsbergen' (1988), the Trust's legendary general director Nikolaj Aleksandrovitsj Gnilorybov presents a thorough description of the development

of mines and settlements that also display the Soviet five-year planning structure and ideology:

> The socialistic competitions, the Stakhanov movement,[48] and elite workers contributed to the successful fulfilment of all planned achievements in 1950 and for the five-year plan in general. At this time a total of 2400 persons were working in our mines. Of these, 1868 persons participated in the competition to carry out our determined tasks within the scheduled time; 1623 persons managed to do so. Of 46 brigades participating in the competitions, 41 managed to fulfil their commitments. The workers' active participation in the socialistic competition facilitated the achievements of results that were even higher than our production norms. The following are some examples of how the determined norms were fulfilled: workers in service facilities – 131.6%, workers in the mines – 136.6%, workers in workshops and on the surface – 136.4%.[49]

The competitions between workers, shifts, brigades and working places, and between personnel and determined plans were subject to a structured system of awards conferred by the Trust as well as the Soviet authorities. These included bonuses, hailing on poster boards (worker of the month), titles of honour and medals, and naming in speeches on important occasions.[50]

In Pyramiden, the Finnish houses (since demolished, but visible on aerial photos from 1960 and 1990), and the first administrative building and a series of stores, workshops and machine houses belongs to the first development of the settlement. Despite the Trust's tremendous efforts in these early years, the first officially registered shipment of coal from Pyramiden was not made before 1956.[51] The coal layers in Pyramiden were a challenge. Unlike the flat and continuous 'cake fill' layers in the Longyearbyen coalfields, the layers in Pyramiden were sloping and interrupted by frequent faults. The Pyramiden coalfields are of Carboniferous age, almost 250 million years older than those in Longyearbyen, and hence more subject to tectonic movements, compression, dipping, and faulting.[52] These geological conditions affected the cost-benefit ratio and were clearly decisive for the mine's close-down in 1998.

The formation of the town square started before 1960, when the two rows of two-floor dwelling apartments were constructed. In the late 1960s, a comprehensive plan for modernizing and mechanizing the mines, and a further development of the settlement was approved by Soviet authorities. The planning involved a series of expert researcher teams. The enterprise aimed at 'securing and increasing the mines' productivity, and improving the service- and cultural facilities for the staff'.[53] This plan was largely fulfilled within the ninth five-year plan (1971–1975). The Trust's general director Gnilorybov could rightfully boast of the fruits of these efforts in his speech for Pyramiden's 30 Year Jubilee on 29 August 1976:

> In the course of 30 years' work the mine Pyramiden has produced almost 4 million tons of coal, and approximately 100 km of the mountain has been surveyed. The external part of the mine has changed beyond recognition. The Soviet people's hands have transformed it to a modern mining society, where all that is needed for work and rest under harsh Arctic conditions is found.

> The settlement counts 32 dwelling complexes. Dwelling facilities are modernized and constantly improved. In 1975, a dwelling complex with 70 apartments was delivered, also containing a café with 75 seats, a complex for daily services, chemical cleaners, a photographic studio, hairdresser, and

workshops for a tailor and shoemaker. A powerful electro central heating has been built. In the bath and laundry complex, in addition to the services prescribed in the main regulation, there is a water treatment facility[54] and two ultra-violet heat radiation apparatuses with a capacity of treating 180 persons per hour. […] In the settlement there is a mess hall with 200 seats, a greenhouse and also houses for animals. In 1975 alone, 35,000 kg of meat was produced, 48,000 litres of milk, 110,000 eggs, and 5700 kg of green onions and other vegetables. For cultural leisure activities, a Cultural Palace, containing a theatre with 370 seats, a gym, a billiard room, and a library with a collection of almost 12,000 books are at the mine collective's disposal. There is also a football stadium named after Jurij Gagarin, a sports arena, a firing range, and six lounges (the so-called 'Red Corners'), and a local museum. An evening school has been established, with preparation courses for applicants to middle and higher educational institutions, and also courses for car and motor bike drivers. The mine's kindergarten is attended yearly by approximately 50 children. In 1975, a primary school for the polar researchers' children was established. Medical services for the mine's workers are handled by a stationary hospital with 20 beds, an ambulance and a pharmacy.

The aims for further improvement of cultural day-to-day conditions for the Soviet polar researchers in Pyramiden in the next few years are to finish the main repairs and rebuilding of the existing dwelling complexes. In each house, there will be one- and two-room apartments with hand washing facilities, built-in wardrobes for outer clothes, and modern furniture. The mess hall is to be rebuilt, production departments will be expanded, and the existing rooms fixed.

Under the tenth five-year plan, the polar researchers will have a new four-floor dwelling complex with 120

apartments, each with a bath, toilet and shower, a new polio clinic in the hospital with 20 beds, and complete sets of all necessary premises and services, and a combined 500 ton depot for vegetables and fruit.[55]

One should bear in mind that this was happening amidst the global schism of cold war, and it is interesting to note that Gnilorybov finished his speech by highlighting the healthy relation between Soviets and Norwegians on Svalbard, praising 'free and unforced' sports competitions and cultural exchange:

> Groups of Longyearbyen inhabitants visit Pyramiden to get to know its people, watch films and exhibitions, and take part in conversations about life in the Soviet Union, about the Soviet peoples' progress in building communism.[56]

The above description of Pyramiden portrays most of the town as it can be seen today. However, the town and mine were developed and modernized until the very end. It seems, however, that resources for maintenance and development were reduced in post-Soviet times. The reduction in shipped coal from Pyramiden from c.1990 and onwards[57] may also have been related to the fall of the Soviet regime. However, there were also problems relating to the productivity of the mine – transport distances increased by the day, and by the end it took one hour to travel from the settlement to the productive coal fields in the mine.[58] The situation was hardly helped by the catastrophic crashing of the Trust's chartered Tupolev in August 1996, killing all 141 Russian and Ukraine passengers and crew members on-board. In addition, one year later, the Barentsburg mine exploded, killing 23 workers.

In 1997, Trust officials started mentioning a possible close-down of Pyramiden in meetings with the Governor of Svalbard, without giving any specific date.[59] Coal reserves were not exhausted, but large investments were needed for developing the mine. The financial situation seemed uncertain, signals from the post-Soviet regime were mixed and the Trust planned to concentrate its enterprise in Barentsburg. Late in 1997, the autumn of 1998 was mentioned as a probable time fore the close-down. Later, everything seemed to accelerate, as the closing down of the mine was moved forward to 1 April 1998.

The last coal from the mine was symbolically placed in a trolley by the Pyramiden monument. All workers were redirected towards dismantling, removing, collecting, moving, and cleaning. The eagerness of removing and cleaning also resulted in the demolishing of the Finnish houses. After this incident, the Governor and the Trust cooperated more closely in deciding what should be removed and how the structures left behind should be protected. The Trust had already decided to secure 'capital buildings' with wooden shutters in front of all windows on the ground floor. The Governor advocated that all that the Trust did not need, and that was not a threat to the environment (e.g. fuel, chemicals) should be left in its context.[60] The last permanent inhabitants left Pyramiden on 10 October 1998. However, Trust Arcticugol employees from Barentsburg frequently visit Pyramiden to collect equipment, and carry out maintenance and the removal/cleaning of hazardous materials.

The last man

I recall my last visit to Pyramiden in March 1999, after the settlement's first winter without its people. The Governor of Svalbard wanted to check how the town had survived. We landed the helicopter in the harbour area, and walked up to the settlement. There were snowdrifts in all the wrong places. No other movement than that from the remnants of a Russian flag on the roof of the administrative building. All of the clocks showed different times. There was no other sound than the uneasy, slow, rhythmic squeaking from children's swings, and occasional various noises from other loose things moving with the wind. No footprints but our own and (fortunately) no signs of polar bear break-ins. Everything seemed all right, but yet all wrong. We did not talk much. Instead we were looking, and worrying about future problems – break-ins, looting, vandalism, souvenir hunting, and the misplaced care for the well-being of things that 'might just be destroyed here – probably best that I bring it home to my window ledge. Saving is not stealing'.

While airborne on returning, we spotted something moving on the ground. It was a human being, running along the road from the frozen power plant, waving his arms. He was the self-appointed German caretaker that had spent the winter in the Trust's cabin in Petunia Bay some kilometres outside the town.

As we landed he had to hold onto his hat to prevent it from blowing away, turning around to avoid the snow-mixed swirls from the rotor. He looked cold in his combat-like jumpsuit and rubber boots, and also lonely from months of isolation. He told us that he had been looking after the town during the winter, closing a few open windows, and hammering some extra nails in loose wooden shutters. He had also paid a visit to his closest neighbour, a trapper just as isolated as himself, by walking the more than 30 kilometre distance. Unfortunately, his unannounced visit had coincided with some unspecified, but important chores that his neighbour was engaged in. However, he was able to have a quick cup of coffee, after which he then walked the same 30 kilometres back to his quarters outside Pyramiden. He pointed at his feet: he had spent 26 hours in these rubber boots.

He had a letter to be delivered, but no stamp or envelope, only a cover made from an extra sheet of paper, all blackened by his soot-stained fingers. Evidently, he did not have glue or tape either, as the edges were closed by 31 paper clips. Taking into account all that he was missing, we were impressed that he had acquired such an amount of paper clips. He asked if we would be so kind as to take the letter to the rightful addressee, and then he had another request, for a cigarette. On being offered a whole pack, he refused: this one was all he needed. We watched him as he smoked it, very slowly. Afterwards we returned for Longyearbyen, together with his letter.

Endnotes

[1] 'The Pyramid' (Norwegian: Pyramiden) – named after the impressive mountain (Mt. Pyramiden) overlooking the site.

[2] The abandonment of Pyramiden was not definitive. At meetings with the Governor of Svalbard in 1998, representatives from Trust Arcticugol informed that there still were coal reserves of commercial value in Pyramiden. But large investments were needed to develop the mine towards new coal layers. Trust Arcticugol was not prepared for such investments in 1998 – and the town was put on hold. This may explain the amounts of seemingly useful equipment and facilities at the site.

[3] For one of us this was a revisit. Working for the Governor of Svalbard in Longyearbyen 1996–1999, Bjerck visited the town several times while it was still occupied and also during its phase of abandonment.

[4] Among the very few exceptions is Liv Balstad's important description of the Governor of Svalbard's summit-like visit to the first Russian Consulate in Pyramiden in 1947 (1955: 326-341). Also of interest are the publications by Trust Arcticugol's former general director, Nikolaj Aleksandrovitsj Gnilorybov (1976, 1979, 1988). Gnilorybov describes the town's technical and physical development in impressive detail, but shows little or no concern for how this impacts on everyday human life. However, he asserts that the 'modern, original architectural plans for buildings and installations that are erected (…) without doubt have a positive influence on peoples' mood' (Gnilorybov 1988: 82). See also the works by Hoel (1965), Gnilorybov and Ivanov (1988), Fløgstad (2007), and Fløgstad and Hermansen (2007).

[5] See Lucas 2004 for a discussion of the concept in relation to the study of the present. The 'prehistoric' in Lucas' conception is not a chronological term (or related to any conceptions of static or 'cold' societies), but designates a field of the non-discursive experiences excluded by (or not yet brought into) discourse.

[6] Phrase borrowed from Serres 1987: 209

[7] Lefebvre 1987: 9

[8] Pyramiden is the property of Trust Arcticugol. Our project description and request to stay in Pyramiden was mediated to Trust Arcticugol through the Governor of Svalbard.

[9] Glassie 1999: 47

[10] Shanks 1997: 102

[11] For projects combining archaeology and art see Coles and Dion 1999, Tilley et al. 2000, Pearson and Shanks 2001, Renfrew 2003, DeMarrais et al. 2004, Witmore 2004. On photography and archaeology, see Shanks 1997.

[12] Arlov 1996: 299-320, see also our section headed Pyramiden in Svalbard, p. 160.

[13] Østreng 1974, Holstmark 1993, Jørgensen 2004

[14] Governor of Svalbard 2009a: *About Svalbard*. Online at [http://www.sysselmannen.no]

[15] The three-room department was hidden behind toilets and store rooms, barricaded by two sets of steel doors and with its windows secured by bars and curtains of steel. The first room appears to have been an office. The second room, lined with soundproof plates, seems to have been a communication centre. A gas-fuelled stove in the inner room was probably used for destroying documents. Collectively, the facilities probably related to cold war strategies.

[16] Gnilorybov 1976

[17] On the aerial photos (p. 164-165) these are represented by the easternmost cluster of buildings diverging from the linear axis.

[18] Augé 1995

[19] See Gonzalés Ruibal 2008 for a different but related case. Regarding Soviet architecture and ideology, see Humphrey 2005; for monuments and ideology, see Lewis 1995, see also Lahusen's interesting paper on socialist ruins (Lahusen 2006).

[20] Edensor 2005: 47

[21] Governor of Svalbard 2009b: *PCB project - 2008 annual report*. Online at [http://www.sysselmannen.no/hoved.aspx?m=55710&amid=2626827]

[22] The term relates to psychologist James Gibson's theory of direct perception, in which landscapes and things have qualities that they offer or afford to us in a direct or unmediated way (Gibson 1979: 127ff.). Here, the concept is used very liberally to convey the portfolio of material expressions associated with 'Sovietness'.

[23] See Miller 1988, 2001, Clarke 2001, Gullestad 2002

[24] See Arkhipov 2006

[25] Often rasterized black & white pictures on photo paper, suggesting some kind of underground distribution of home-made copies from printed magazines.

[26] See Lucas 2002, Shanks et al. 2004, Scanlan 2005

[27] Edensor 2005: 22, 95

[28] Edensor 2005: 139

[29] Edensor 2005: 117

[30] DeSilvey 2006

[31] Benjamin 1999, Anderson 2001

[32] DeSilvey 2006

[33] Edensor 2005: 109-110

[34] See Cresswell 1997: 335

[35] Benjamin 1985: 162-66, 2003: 169
[36] Benjamin 1999: 473-476, Buck-Morss 1999: 110ff., 220-21
[37] Buck-Morss 1999: 219
[38] Benjamin 1999: 13
[39] Information on legislation and government in this chapter is from Governor of Svalbard 2009c: *Legislation*. Online at [http://www.sysselmannen.no]
[40] Ministry of the Environment 2001. Online at [Government.no]
[41] Gnilorybov 1988: 13
[42] Gnilorybov 1988: 20ff.
[43] Gnilorybov 1976: 3
[44] Gnilorybov 1988: 28
[45] Gnilorybov 1976: 5
[46] Balstad 1955: 332ff.
[47] I.e., Pyramiden, Grumant and Barentsburg, Gnilorybov 1988: 40
[48] A Soviet workers' movement for productivity increase and efficiency, named after the legendary mine worker A.G. Stakhanov.
[49] Gnilorybov 1988: 40, our translation
[50] E.g., the 22 personal awards included in Gnilorybov's speech (1976: 12ff.).
[51] Personal communication with Torfinn Kjærnet, senior engineer, The Directorate of Mining with the Commissioner of Mines at Svalbard. However, Gnilorybov (1976: 8) states that Pyramiden shipped considerable amounts of coal to Murmansk and Arkhangelsk from 1948.
[52] Gnilorybov 1988: 58ff., and personal communication with Torfinn Kjærnet, senior engineer, The Directorate of Mining with the Commissioner of Mines at Svalbard.
[53] Gnilorybov 1988: 44, our translation
[54] Large tubs of hot water
[55] Gnilorybov 1976: 8ff., our translation
[56] Gnilorybov 1976: 15, our translation
[57] As registered in the official archives of The Directorate of Mining with the Commissioner of Mines at Svalbard, personal communication with Torfinn Kjærnet.
[58] Personal communication with Torfinn Kjærnet.
[59] As a member of the Governor's staff, Bjerck took part in these meetings, and also assisted in implementing cultural heritage concerns (protecting and securing buildings, structures, and artefacts, removal of environmental hazards, etc.) during the process of the abandonment of Pyramiden.
[60] Letter from the Governor of Svalbard to Trust Arcticugol, 18 May 1998

Bibliography

Andersson, Dag T. (2001) *Tingenes taushet, tingenes tale.* Oslo: Solum.

Arkhipov, Vladimir (2006) *Home-Made. Contemporary Russian Folk Artifacts.* London: Fuel.

Arlov, Thor B. (1996) *Svalbards historie.* Oslo: Aschehoug.

Augé, Marc (1995) *Non-Places. Introduction to the Anthropology of Supermodernity.* London: Verso.

Balstad, Liv (1955) *Nord for det øde hav.* Bergen: J. W. Eides Forlag.

Benjamin, Walter (1985) *The Origin of the German Tragic Drama.* London: Verso.

Benjamin, Walter (1999) *The Arcades Project.* Cambridge, Massachusetts: The Belknap Press of Harvard University Press.

Benjamin, Walter (2003) *Selected Writings Volume 4: 1938–1940.* Cambridge, Massachusetts: The Belknap Press of Harvard University Press.

Buck-Morss, Susan (1999) *The Dialectics of Seeing. Walter Benjamin and the Arcades Project.* Cambridge, Massachusetts: The MIT Press.

Clarke, Alison J. (2001) The aesthetics of social aspiration. In Daniel Miller (ed.), *Home Possessions. Material Culture behind Closed Doors.* New York: Berg Publishers.

Coles, Alex and Mark Dion (eds) (1999) *Archaeology: Mark Dion.* London: Black Dog Publishing.

Cresswell, Tim (1997) Weeds, plagues, and bodily secretions: a geographical interpretation of metaphors of displacement. *Annals of the Association of American Geographers* 87 (2): 330-345.

DeMarrais, Elisabeth, Chris Gosden and Colin Renfrew (eds) (2004) *Substance, memory, display: Archaeology and art.* Cambridge: McDonald Institute.

DeSilvey, Cairlin (2006) Observed decay: Telling stories with mutable things. *Journal of Material Culture* 11 (3): 318-338.

Edensor, Tim (2005) *Industrial Ruins. Space, Aesthetics and Materiality.* New York: Berg.

Fløgstad, Kjartan (2007) *Pyramiden. Portrett av ein forlaten utopi.* Oslo: Spartacus.

Fløgstad, Kjartan and Siri Hermansen (2007) Sub sole. *Syn og Segn* 0107: 44-59.

Gibson, James (1979) *The ecological approach to visual perception.* Hillsdale NJ: Lawrence Erlbaum Associates

Glassie, Henry (1999) *Material culture.* Bloomington and Indianapolis, IN: Indiana University Press.

Gnilorybov, Nikolaj Aleksandrovitsj (1976) «Пирамида» – Советский угольный рудник на архипелаге Шпицберген (1946–1976 гг.). ['Pyramiden' – A Soviet Coal Mine in the Spitsbergen Archipelago (1946–1976)]. Speech given by Trust Arcticugol's general director Nikolaj Aleksandrovitsj Gnilorybov at Pyramiden's 30th anniversary. Unpublished manuscript, translated into Norwegian by the Governor of Svalbard. Longyearbyen: Archives of the Governor of Svalbard.

Gnilorybov, Nikolaj Aleksandrovitsj (1979) *Советский угольный рудник «Пирамида» на архипелаге Шпицберген.* [The Soviet Coal Mine 'Pyramiden' in the Spitsbergen Archipelago.] Москва: ЦНИЭИуголь.

Gnilorybov, Nikolaj Aleksandrovitsj (1988) *Угольные шахты на Шпицбергене.* Москва: Недра. [The Coal Mines at Spitsbergen] Unpublished Norwegian translation by T. Bøhler for The Governor of Svalbard. Longyearbyen: Archives of the Governor of Svalbard.

Gnilorybov, Nikolaj Aleksandrovitsj and G. Ivanov (1988) *ПОД НЕБОМ АРКТИКИ.* [Under polarhimmelen.] МОСКВА: ПЛАНЕТА.

Gnilorybov, Nikolaj Aleksandrovitsj and G. Ivanov (1989) *Under polarhimmelen.* [Norwegian translation of text in Gnilorybov & Ivanov (1988), included as an addendum to the book.] Longyearbyen: Svalbardpostens trykkeri.

González-Ruibal, Alfredo (2008) Time to Destroy: An Archaeology of Supermodernity. *Current Anthropology* 49: 247-263.

Governor of Svalbard (2009a) *About Svalbard*. Online at [http://www.sysselmannen.no]

Governor of Svalbard (2009b) *Pollution: PCB project - 2008 annual report*. Online at [http://www.sysselmannen.no/hoved.aspx?m=55710&amid=2626827]

Governor of Svalbard (2009c) *Legislation*. Online at [http://www.sysselmannen.no]

Gullestad, Marianne (2002) *Kitchen-table society*. Oxford: Oxford University Press.

Hoel, Adolf (1965) *Svalbards historie*, bd. 1. Oslo: Sverre Kildals Boktrykkeri.

Holstmark, Sven G. (1993) *A Soviet Grab for the High North? USSR, Svalbard and Northern Norway 1920–1953*. Forsvarsstudier 7/1993. Oslo.

Humphrey, Caroline (2005) Ideology in Infrastructure: Architecture and Soviet Imagination. *Journal of the Royal Anthropological Institute* 11 (1): 39-58.

Jørgensen, Jørgen Holten (2004) Svalbard: russiske persepsjoner og politikkutforming. *Internasjonal politikk* 62 (2): 177-197.

Lahusen, Thomas (2006) Decay or endurance? The ruins of socialism. *Slavic Review* 65 (4): 736-746.

Lefebvre, Henri (1987) The everyday and everydayness. *Yale French Studies* 73 (fall): 7-11.

Lewis, Mark (1995) On the Monuments of the Republic. *October* 73, 1995.

Lucas, Gavin (2004) Modern Disturbances. On the Ambiguities of Archaeology. *Modernism/modernity* 11: 109-20.

Miller, Daniel (1988) Appropriating the state on the council estate. *MAN* 23 (2): 353-72

Miller, Daniel (2001) Possessions. In Daniel Miller (ed.), *Home Possessions. Material Culture Behind Closed Doors*. New York: Berg Publishers.

Ministry of the Environment (2001) *Act of 15 June 2001 No.79 relating to the protection of the environment in Svalbard*. Online at [Government.no]

Pearson, Mike and Michael Shanks (2001) *Theatre/Archaeology*. London: Routledge.

Renfrew, Colin (2003) *Figuring it out: What are we? Where do we come from? The parallel visions of artists and archaeologists*. London: Thames & Hudson.

Scanlan, John (2005) *On garbage*. London: Reaktion Books.

Serres, Michel (1987) *Statues*. Bourin, Paris.

Shanks, Michael (1997) Photography and archaeology. In Brian Molyneaux (ed.), *The cultural life of images*. London: Routledge.

Shanks, Michael, David Platt and William L. Rathje (2004) The perfume of garbage. In *Modernity/Modernism*, 11 (1): 68-83.

Tilley, Chris, Sue Hamilton and Barbara Bender (2000) Art and the re-presentation of the past. *Journal of the Royal Anthropological Institute* (n.s.), 6 (1): 36-62.

Witmore, Christopher (2004) Four archaeological engagements with place: mediating bodily experience through peripatetic video. *Visual Anthropology Review*, 20 (2): 157-72.

Østreng, Willy (1974) *Økonomi og politisk suverenitet. Interessespillet om Svalbards politiske status*. Oslo: Universitetsforlaget.

Photo index

Page 3. Arrangement in the Cultural Palace. Photo found in Cultural Palace. Photo E. Andreassen

Page 17. Elin documenting in the mechanical workshop. Note the cover of white dust from crumbling paint and plaster. Photo H. Bjerck

Page 4. Winter festival at playground during Pyramiden's heyday. The 'Finish houses' in background. Photo found in Cultural Palace. Photo E. Andreassen

Page 18. Decorative curtain of coloured plastic pipes on cotton thread in the entrance to the main room in worker apartment in block 38. Photo E. Andreassen

Page 8. The bar in Hotel Tulip. Note the three-dimensional wall designs – padded red dimond quilt fabric – and the alternating plastic covered modules to the right. (see p. 78). Photo E. Andreassen

Page 20. Decorative curtain made of used typewriter ribbons in door between hall and main room, block 38. Photo E. Andreassen

Page 12. The block for higher staff, and remnants from the former family society in Pyramiden (see p. 4). Metal boxes in windows are natural 'refrigerators' that may be opened from inside. New residents have taken over in very orderly fashion, with three nesting gulls on each ledge. Stains on the wall reveal severe water leakage. (see pp. 138, 140). Photo E. Andreassen

Page 22. From the museum in Hotel Tulip. To the far right, duct tape mended window after the second shooting of the stuffed polar bear. Photo E. Andreassen

Page 14. Dried out plants in boxes made of orange juice cartons, and a decorative foam rubber fox (see p. 126). View from apartment in block 38, overlooking the low part of the town square. Note freezing device by the block in centre (white poles), ensuring stable ground during summer thawing. Photo E. Andreassen

Page 25. Administrative building in harbour area. Photo E. Andreassen

Page 26. Decoration of cigarette cartons and boxes in worker apartment in block 38. Home-made book shelf, chair mended with woollen blanket, and a lamp shade on the table. Photo E. Andreassen

Page 36. The mechanical workshops – drawers still open after the final check for usable parts. Photo E. Andreassen

Page 29. From room for resting on duty (?) in the garage. Photo E. Andreassen

Page 38. The garage. Photo E. Andreassen

Page 30. Handmade dumb-bells (and a baggage tag) on linoleum-tiled floor, apartment in block 38. Photo H. Bjerck

Page 41. Workshop in garage. Photo E. Andreassen

Page 32. Workshop in the garage, windows blocked by shutters put up before the abandonment of Pyramiden in 1998. Photo E. Andreassen

Page 42. Mechanical workshop. Photo E. Andreassen

Page 34. The control room at the coal-fuelled power plant. Photo E. Andreassen

Page 45. Lenin monument in front of the Cultural Palace. Part of stand for flags at bottom right, see p. 170. Photo B. Olsen

Page 46. The cemetery. The 43 graves are mainly from the late 1940s and 1950s – one of the graves has been exhumed, probably during the abandonment of the settlement. Pyramiden and Mt. Pyramiden in background. Photo E. Andreassen

Page 56. Communication tower at the heliport. Photo E. Andreassen

Page 48. View from the cemetery, overlooking Billefjorden and Nordenskjöld Glacier. Photo E. Andreassen

Page 59. Communal room, block 38. These larger rooms, used for spare-time activities, are found in each floor of the apartment complex. Photo H. Bjerck

Page 50. Pipeline cage walkway between the power plant and the town centre – part of 'The Street for the 60 Year Jubilee of the Great October' (see p. 166). This photo also shows the unstable soils at the base of Mt. Pyramiden. Photo H. Bjerck

Page 60. Covered walk way from mine bath to cable car/mine entrance. Note the absence of graffiti on the walls. Photo E. Andreassen

Page 53. Twin peaks – the initial administrative building and the impressive summit of Mt. Pyramiden. The covered cable car and coal conveyor belt from the mine to the left. Photo E. Andreassen

Page 62. Sunflower monument (and flag socket) in the town square, marking a communal grave for cats. Also note the specially cultivated Arctic grass. Photo E. Andreassen

Page 54. The Lenin statue overlooking the flooded town square in Pyramiden. View from the entrance to the Cultural Palace. Photo E. Andreassen

Page 65. Pipeline cage to the Blue Lagoon water reservoir. The active and free flowing meltwater has avoided the bridge and regained its natural course, removing the road in the process. Dikes and bulldozer-deepened water channels were needed to control water flow during the settlement, cf. remnant of dike to the right of the bridge. Photo H. Bjerck

Page 66. Nesting gulls on window ledges in the waterworks. Photo H. Bjerck

Page 77. Bar (left) and dining room on the first floor of block for higher staff – also used for entertaining official visitors. Photo E. Andreassen

Page 68. Hein and Bjørnar preparing dinner in our suite at the top floor of Hotel Tulip. Photo E. Andreassen

Page 79. Behind the bar in Hotel Tulip. Flowers of straw and paper clippings, illustrating both Pyramiden creativity and the length of the imported grass in the town square (cf. page 62). Photo E. Andreassen

Page 70. Hotel Tulip, with museum, bar and post office on ground floor. Photo E. Andreassen

Page 80. The mosaic in mess hall, Pyramiden's initial Cultural Palace. Photo E. Andreassen

Page 73. Lunch break on the airy and safe roof of Hotel Tulip. Photo E. Andreassen

Page 83. Lounge in library, Cultural Palace. Photo E. Andreassen

Page 74. The theatre in the Cultural Palace. Photo B. Olsen

Page 84. From one of the polar researchers' offices in administrative building. Photo E. Andreassen

Page 87. Waste deposit in Pyramiden – mainly ash from the coal-fuelled power plant. Photo E. Andreassen

Page 97. Apartment in block 38. Photo E. Andreassen

Page 88. Shelf (for water heater?), apartment in block 38. A simple, but functional example of a wide variety of home-made shelves. Also note one of many missing electrical switches and sockets – apparently systematically removed during the process of leaving Pyramiden. Photo E. Andreassen

Page 98. Apartment for higher staff. Photo E. Andreassen

Page 91. Entrance to a home in block 38 – that also contained the glass painting (p. 112) and the 'Brooklyn' decoration (p. 26). Photo B. Olsen

Page 101. Apartment in block 38. Photo E. Andreassen

Page 92. Bedroom in apartment for higher staff. Photo E. Andreassen

Page 103. Apartment in block 38. Photo E. Andreassen

Page 95. From apartment in block 38. Photo E. Andreassen

Page 104. Apartment in block 38. Photo E. Andreassen

Page 107. Apartment in block 38. Photo E. Andreassen

Page 117. Remains of erotic black and white pictures, inside wardrobe doors in apartment in block 38. Rasterized black and white pictures on photo paper suggest some kind of underground distribution of home-made copies from printed magazines. Photo E. Andreassen

Page 109. Apartment in block 38. Photo E. Andreassen

Page 118. A collection of politicians' portraits, on inside of wardrobe door. Apartment in block 38. Note remains of extra locking devices. Photo E. Andreassen

Page 110. The remains of a home in block 38. An armchair (with one new leg), between a home-made bookshelf and a corner shelf (for radio or cassette player?), a wall-mounted ceiling neon light. Collage dominated by paper clippings of cars. Photo E. Andreassen

Page 120. Karl Marx portrait and antenna with clothes peg in apartment in block 38. Photo E. Andreassen

Page 112. Glass painting with exotic details on inner door of apartment in block 38. Collage of cigarette boxes to the left, poster of women in the middle. Photo E. Andreassen

Page 122. Flower arrangement in apartment, block 38. Shelf made of thin boards from wooden boxes, decorated with antenna cable, and an illuminated central compartment. Flowerpots made of recycled lamp shades. The thin plate (barely visible) below the white table is part of a hidden compartment. Photo E. Andreassen

Page 115. Pull-up bar and weight lifting bench in apartment for higher staff. Photo E. Andreassen

Page 125. Decorative mobile of paper clippings hanging in window above the radiator in dwelling, block 38. Photo E. Andreassen

Page 126. Home-made foam rubber fox in chain made of paper clips and empty tape roll, apartment in block 38 (see p. 14). Photo E. Andreassen

Page 136. Blurred view of the town square. From top floor window in the Cultural Palace. Photo B. Olsen

Page 128. Wall decoration, apartment in block 38. Photo E. Andreassen

Page 139. Bedroom in apartment for higher staff, severely damaged by water leakage (see p. 12). Photo E. Andreassen

Page 130. Wall decorations, worker apartment in block 38. At centre, an example of Trust's 4-year calendars marking days, weeks and months spent in Pyramiden, reflecting the standard 2-year staff contracts (see p. 104). Here, a uniformed man appears to be pointing at a date of departure (?). Photo E. Andreassen

Page 140. Meltwater dividing the playground by the block for higher staff, undermining and destabilising the building, causing cracks and severe leakage (see pp. 12 and 138). First floor window shutters are not enough to protect the building – attacks always come from unexpected directions. Photo E. Andreassen

Page 133. Apartment in block for higher staff. The large pot plant, wallpaper poster and radiator once formed a micro-environment that is hard to find in the barren Arctic. Note the wallpaper on the ceiling, and the blue lining (clippings from a different kind of wallpaper) between ceiling and wall. Photo B. Olsen

Page 145. Lamps and mirrors removed from apartments (presumably for further use elsewhere), apparently as part of the closing down plan. However, in block 38, most of the items ended up in a room in the first floor. Photo. E. Andreassen

Page 135. Locally produced TV/video player rack and wall paper poster in apartment for higher staff. Note the general absence of individual decorations in these apartments (e.g. pp. 92, 115, 133, 150). In foreground, 'wallpaper' fallen from ceiling. Photo E. Andreassen

Page 146. Rubber boots in storage room for butchered pigs – door to slaughtering room (p. 148) to the left. Photo E. Andreassen

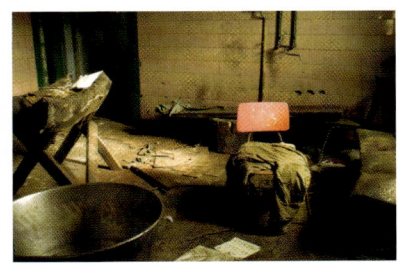

Page 148. Facilities in the slaughtering room in the pig house. Stained, metal-lined butchering bench to the left, the cut-off electrocuting instrument on the floor. Photo E. Andreassen

Page 158. The slogan (made of wooden beams) on the steep slope facing Pyramiden. The original 'hammer and sickle' motif no longer exists, but 'Miru Mir' (Peace on Earth) and the later addition of 'Tula', one of the Russian home towns, are still visible. The mine entrance and covered cable car are to the right. Photo B. Olsen

Page 150. Apartment for higher staff with door damaged by a break-in. Dead seagull under sofa. Photo E. Andreassen

Page 161. Framed photo of the Norwegian King and Queen's visit to Pyramiden in 1995 (left) and a stuffed Svalbard reindeer (R. tarandus platyrhynchus) at the Pyramiden museum in Hotel Tulip. Photo E. Andreassen

Page 153. Shoulder (?) padding and medicine bottle on window ledge of apartment for higher staff. Photo B. Olsen

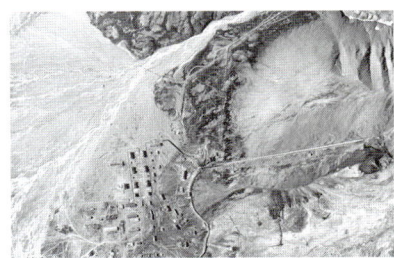

Page 164. Pyramiden and surroundings in 1960. Part of aerial photo S60-7161, printed by permission from the Norwegian Polar Institute ©

Page 154. Home-made device for doing sit-ups, apartment in block 38. Photo H. Bjerck

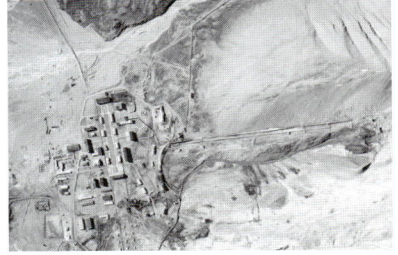

Page 164. Pyramiden and surroundings in 1990. Note the latest expansion into the active river plain, and the artificial dikes and channels directing the meltwater flow towards the bridge (see p. 64). Part of aerial photo S90-1556, printed by permission from the Norwegian Polar Institute ©

Page 156. Detail from the city dump – a child's dummy. Photo H. Bjerck

Page 166. Central part of Pyramiden in 1990. Part of aerial photo S90-1556, printed by permission from the Norwegian Polar Institute ©

 Page 170. Star and 'hammer and sickle' decorated stand for banners and flags in the town square in Pyramiden, looking towards the Cultural Palace. All lamp posts have small crowns with 5 flags sockets. In all, 242 sockets for different types of flags and banners are located in and around the town square. Photo E. Andreassen

 Page 181. Leonid Bresjnev poster, decorated with additional paper medals. Apartment in block 38. Photo B. Olsen.

 Page 172. Music studio in the Cultural Palace. Photo B. Olsen

 Page 182. The mess hall in Pyramiden, prior to 1988. From Gnirolybov & Ivanov (1988), p. 110. Photo V. Ivanov & A. Sergejev

 Page 174. Storage for mining equipment, all with a protective coating of grease and shipshape. Trust Arcticugol regularly collect items stored in Pyramiden for use in Barentsburg. Photo E. Andreassen

 Page 185. Dead snow bunting (Plectrophenax nivalis) on desk in office, administrative block. Photo H. Bjerck

 Page 176. Smaller dining room in mess building, the initial Cultural Palace in Pyramiden. Wall lined with mosaic of blue/white stencil painted glass plates. Photo E. Andreassen

 Page 188. Paper flowers on imported twigs in vase made of plastic bottle filled with plaster for stability. Reception desk in store house for work clothes. Photo E. Andreassen

 Page 179. Portrait photo, on table in apartment, block 38. Photo E. Andreassen

 Page 192. Pyramiden and Billefjorden from the slope of Mt. Pyramiden. Photo B. Olsen

Page 206. Handmade electrical barbecue, harbour area: a stand made of welded iron rods, electrical resistors around a ceramic cylinder (not recommended for indoor use). Photo E. Andreassen

Page 208. Interior of the bottle house, situated north of the town, used for social activities and recreation. The house is built of c. 5300 bottles (our calculation), fixed in concrete on a floating steel construction to prevent damage caused by freeze-thaw cycles. Photo E. Andreassen

Page 211. Two generations of 'Emergency exit' signs at Hotel Tulip. The newest is a hand-made copy of the signs prescribed by the Norwegian authorities. Letters and lines have been cut from adhesive fabric and glued onto green-painted squares. Photo H. Bjerck

Elin Andreassen (1965), educated from Bergen National Academy of the Arts, Norway (1998). Her work is rooted in social relations and power structures, often related to her North Norwegian background. Beside individual works, she also develops projects in colaboration with other artists, historians, writers and musicians.

Hein B. Bjerck (1954), associate professor of archaeology, Museum of Natural History and Archaeology, Norwegian University of Science and Technology (NTNU). Cultural Heritage Officer at the Governor of Svalbard's Department of Environment Conservation (1996–1999).

Bjørnar Olsen (1958), professor in archaeology at the Institute of archaeology and social anthropology, University of Tromsø, Norway. His research interests include material culture studies, theoretical archaeology, museology, Sámi history/archaeology, Northern prehistory and early history.

Acknowledgements

Commissioner of Mines at Svalbard
Governor of Svalbard
Institute for Archaeology and Sosial Anthropology, University of Tromsø
Museum of Natural History and Archaeology, Norwegian University of Science and Technology (NTNU)
Stanford Archaeology Center, Stanford University
Svalbard Museum
The Norwegian Polar Institute
Trust Arcticugol

Birger Amundsen
Thor Bjørn Arlov
Douglass W. Bailey
Rune Bergstrøm
Reidar Bertelsen
Terje Brattli
Axel Christophersen
Alfredo Gonzáles-Ruibal
Tora Hultgreen
Jørgen Holten Jørgensen
Torfinn Kjærnet
Torbjørn Lefstad
Vladimir Makarov
Ole Jacob Malmo
Olga Mirzaeva
Boris Josifovitsj Nagajuk
Mari Nygård
Per Kyrre Reymert
Mari Røstvold
Jon Ove Scheie
Michael Shanks
Annicken Stuler
Catriona Turner
Aleksander Veselov
Tim Webmoor
Chris Witmore
Andrej Zaremba
Ole Kristian Øye
Harald Faste Aas

© Tapir Academic Press
Trondheim, Norway 2010

ISBN 978-82-519-2436-8

© photo
Elin Andreassen
Hein B. Bjerck
Bjørnar Olsen
© text
Hein B. Bjerck
Bjørnar Olsen

Graphic design:
Elin Andreassen

Cover design:
Klipp & Lim as

Typefaces:
Garamond

Paper:
Arctic matt 150g

Printing:
07 Gruppen as

Published by
Tapir Academic Press
7005 Trondheim, Norway
Tlf.: (+ 47) 73 59 32 10
E-post: post@tapirforlag.no
www.tapirforlag.no
Forlagsredaktør:
mari.nygard@tapirforlag.no

All rights reserved. No part of this publication may be reproduced or transmitted in any form or by any means without prior permission in writing from the publisher.

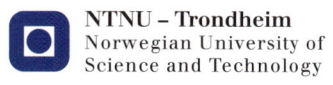